Maths for the Building Trades

Maths for the
Building Trades

JAMES KIDD & IAN BELL

LONGMAN

Addison Wesley Longman Limited
Edinburgh Gate, Harlow
Essex CM20 2JE, England
and Associated Companies throughout the world

© Addison Wesley Longman Limited 1997

First published 1997

British Library Cataloguing in Publication Data
A catalogue entry for this title is available from the British Library

ISBN 0-582-29491-6

Set by 25 in Times 10/12 and Univer
Transferred to digital print on demand, 2002
Printed and bound by Antony Rowe Ltd, Eastbourne

Contents

Preface

Many people who start working in the building industry slowly realise how ill prepared mathematically they are for the trade they have chosen. This is usually discovered when their work colleagues ask them such everyday questions as 'How much materials do you want to order?', 'What's the area of it?' and 'How much will it cost?'

Until recently in the industry it was thought fashionable to boast about being 'no good at maths', but such an admission today will often bar the person from progression within his or her chosen occupation. The skilled operative can be, and in all probability will be, expected almost on a daily basis to have reason to practise mathematical skills, for example:

- Carpenters are required to find the angles of cuts and bevels for pitched roof construction, the height and number of rises in a staircase and the areas of floors and panelling.
- Bricklayers: the number of bricks, the amounts and proportions of sand and cement required. The setting out of building lines, arches and angles.
- Plumbers: the areas and volumes of cisterns and cylinders, the length of runs and carrying capacities of gutters and pipes.
- Heating engineers: the sizes and outputs of radiators, boilers and circulators.
- Painters and decorators: the amount of paint and paper to order.
- If you are self-employed or work for a smaller firm you will in all probability be expected to 'take off' and cost material and labour.

All of these skills – and, of course, there are many more – require the use of mathematics in the form of addition, subtraction, multiplication and division.

Every worker employed in the construction industry requires a certain level of mathematical skills. The level will rise and fall depending on the job they do or the type of person they are. There is, however, a solid core of mathematical knowledge that must be acquired to enable anyone to function as a skilled member of a construction team, and also as a basis for future progress.

Mathematics can be a difficult subject to master, and for many people their first introduction to the subject often left them with painful memories.

There are many reasons for this: it may have seemed at the time to have very little purpose, given the world they were then surrounded by, and the textbooks being used rarely gave detailed explanations that could be readily understood, or perhaps they tried to cover too many areas.

In recent years the arrival of the electronic calculator has removed some of the stress that can be felt when faced with a mathematical problem and in fact calculators have become almost an indispensable part of the trades person's tool kit, and their use is encouraged throughout this book. However, with their continued use it is all too easy to forget, or indeed never really experience, how the system of numbers works, and this is what *Maths for the Building Trades* sets out to explain. Wherever possible, in its explanation of mathematics, it uses everyday problems encountered in the industry by all the various trades involved.

Mathematics plays an important part in our everyday lives whether it be at work or at home. It helps with our understanding of the world in which we live; it makes sure buildings are straight and puts men on the Moon. It is not always a simple subject to grasp and will require hard work on your part. You should, however, gain a valuable skill that will assist and help you throughout your working life and beyond.

James Kidd
1997

How to use this book

The book is divided into chapters or units of study. Each unit is self-contained and can be read on its own if required. The book is fully cross-referenced where needed, so that you can easily look back to chapters that relate to the area being studied. At the end of most chapters is a series of self-assessment questions which are designed to enable you to assess your understanding of the subject just covered. If, when you look at the topics covered in a chapter, you feel you are already skilled in the subjects mentioned, try the self-assessment questions. If you have any difficulty in answering them, work through the chapter.

However, to obtain maximum benefit, and unless you are really confident of the subject, you would be well advised to read and study the book in the order in which it is laid out.

Exercises are given at the appropriate place throughout each chapter and you will reap the greatest benefit by working through them as they occur.

The speed and the rate at which you learn is up to you. I would suggest that you set aside a period of time, maybe an hour each day, for your study, making sure you are confident in that area before moving on to the next.

CHAPTER 1 A few basic rules

Although the use of a calculator is encouraged whilst studying with this book, you should always be able to check the results your calculator gives you and perhaps what is more important fully understand how you have obtained them.

With this in mind you would be well advised to take a few moments to familiarise yourself with the contents of this chapter. It contains many of the rules of mathematics that we shall be using as we work through the book. Most of them you will probably already know and understand. Nevertheless, please check that they actually mean what you thought they meant.

1.1 Numbers

Counting involves the use of only ten symbols or digits (0, 1, 2, 3, 4, 5, 6, 7, 8, 9), and a system of place value. It is a decimal system and as such its fundamental grouping, known as the base, is ten.

Look at the number **1724832**. Each digit has a place value and the location of that digit indicates its value. Starting from the right-hand side of the number, the digits represent ones, tens, hundreds, thousands, tens of thousands, hundreds of thousands and thousands of thousands, or millions.

Example

The number 1724832 should be thought of as

2 lots of one
3 lots of ten
8 lots of one hundred
4 lots of one thousand

2 lots of ten thousand
7 lots of one hundred thousand
1 lot of one million

or

$1,000,000 + 700,000 + 20,000 + 4,000 + 800 + 30 + 2$

Note: It is usual in long numbers to separate the digits into groups of three, divided by commas. Starting from the right, 1724832 becomes 1,724,832 and is read as 'one million, seven hundred and twenty-four thousand, eight hundred and thirty-two'.

 Exercise 1.1

Write down in words the following numbers

(a)   67,482
(b)   234,567
(c)   856.433
(d)   4,467,386

1.2 Mathematical symbols

Symbols are used in mathematics as a kind of shorthand. You are probably already familiar with the symbols for addition ($+$), subtraction ($-$), multiplication ($\times$) and division ($\div$), and below are some more symbols and their meanings.

$=$	is equal to
$\neq$	is not equal to
$\approx$	is approximately equal to
$<$	is less than
$\leqslant$	is less than or equal to
$>$	is greater than
$\geqslant$	is greater than or equal to
$\%$	percent
$\propto$	is proportional to
$:$	ratio

Brackets or parentheses, ( ) or [ ] or { }, are used to group operations together.

1.3 Mathematical definitions

The answer obtained when numbers are added together is called the **sum**. Numbers can be added together in any order; for example, adding together the numbers $2 + 4 + 6$ will give the same answer (12) as $4 + 6 + 2$ or $6 + 4 + 2$.

The process of taking one number away from another is called subtraction. The result obtained is called the **difference**.

Example

What is the *difference* in length between two scaffold poles, one 6 metres long and the other 4 metres long?

Subtracting 4 metres from 6 metres gives a difference of 2 metres.

The order of the numbers to be subtracted is important — for example, $6 - 4$ is not the same as $4 - 6$.

When two numbers are multiplied the result obtained is called the **product** of those two numbers.

Example

If three carpenters each worked for eight hours on the same job, how many hours in total will have been worked, or what is the product of 3 multiplied by 8?

The order in which numbers are multiplied is not important; 3×8 will give the same result (24) as 8×3.

When two numbers are divided by each other the answer obtained is called the **quotient**. The number we are dividing by is known as the **divisor** and the number we are dividing into is the **dividend**. The order in which numbers are divided is important: $10 \div 2$ is not the same as $2 \div 10$.

1.4   Order of mathematical operations

Often when carrying out calculations you may find you need to do more than one mathematical operation, or even combine several operations such as addition, subtraction, division and multiplication, to obtain the answer required. For example, working out the total cost of the various lengths and sizes of timber required for a job can involve addition, multiplication and in some cases division and subtraction.

The multiplication or division in the calculation must be carried out before any addition and subtraction is attempted. Multiplication and division have equal priority, as do addition and subtraction.

For example, the calculation $4 + 3 \times 9$ is really asking you to multiply three by nine and add four to the product, i.e.

$$4 + 3 \times 9 = 4 + 27 = 31$$

Similarly, the calculation

$$6 \times 4 - 15 \div 3 + 2$$

would become

$$24 - 5 + 2 = 21$$

where the multiplication and division have been carried out first followed by the subtraction and addition. The majority of calculators have this ability programmed into them as standard algebraic rules. See Chapter 2 for more information.

1.5 Brackets

The use of **brackets** in the example above would have made it much clearer, i.e.

$$(6 \times 4) - (15 \div 3) + 2 = 24 - 5 + 2 = 21$$

Brackets should always be used where there is a chance of ambiguity. The rules of brackets are quite simple:

1. First, work out the contents of the brackets.
2. Next, do any multiplication and division (in order from left to right).
3. Last, do all additions and subtractions (in order from left to right).

Example

$(12 - 4) \times (4 + 3) - 16 \div 4 \times 2$	Work out the contents of the brackets.
$= 8 \times 7 - 16 \div 4 \times 2$	Do all multiplication and division
$= 56 - 16 \div 8$	(in order left to right).
$= 56 - 2$	Do all additions and subtractions
$= 54$	(in order from left to right).

When the contents of brackets are to be multiplied together the multiplication sign is often excluded, for example:

$$(12 - 4)(4 + 3) \quad \text{means the same as} \quad (12 - 4) \times (4 + 3).$$

1.6 Factors

The factors of a number are those numbers that will divide exactly into it; for example, the factors of 20 are 1, 2, 4, 5, 10 and 20.

$$20 \div 1 = 20$$
$$20 \div 2 = 10$$
$$20 \div 4 = 5$$
$$20 \div 5 = 4$$
$$20 \div 10 = 2$$
$$20 \div 20 = 1$$

1.7   Prime numbers

A prime number is a number that cannot be divided by any number except one and itself.

The number 17 is a prime number; the only numbers it can be divided by are 1 and 17. The numbers 1, 2, 3, 5, 7, 11, 13, 17 and 19 are all prime numbers.

 Exercise 1.2

What are the next five prime numbers in the above sequence?

1.8   Common factors

A common factor is, as its name suggests, a factor or a number that is common to two or more numbers. For example, 6 is a common factor of 12 and 36 as $6 \times 2 = 12$ and $6 \times 6 = 36$.

 Exercise 1.3

What is a common factor of 9 and 27?

1.9   Highest common factor

The highest common factor (HCF) of a set of numbers is the largest whole number that will divide into the numbers exactly. Although 3 is a common factor of 18, 36, 45 and 81, it is not their highest common factor.

Exercise 1.4

Find the HCF of the numbers, 18, 36, 45 and 81.

1.10   Lowest common multiple

The lowest common multiple (LCM) is the smallest number that is a common factor of two or more numbers; for example, the lowest common multiple of 3, 4, 6 and 8 is 24. This is the lowest number that 3, 4, 6 and 8 will divide into exactly.

Answers to exercises

1.1 (a) Sixty-seven thousand, four hundred and eighty-two.
(b) Two hundred and thirty-four thousand, five hundred and sixty-seven.
(c) Eight hundred and fifty-six thousand, four hundred and thirty-three.
(d) Four million, four hundred and sixty-seven thousand, three hundred and eighty-six.

1.2 The next five prime numbers in the sequence are 23, 29, 31, 37 and 41.

1.3 3 and 9 are both common factors of 9 and 27.

1.4 9 is the HCF of 18, 36, 45 and 81.

Calculators

2.1 Ready reckoner

At one time the biggest problem that faced those having difficulties with mathematics was the need to recall the times tables. This combined with remembering the rules of mathematics used up much time and effort. Now, of course, there is a machine that can take some of the worries of mathematics away: the electronic calculator. If you do not always carry a calculator keep a ready reckoner (Figure 2.1) with you at all times.

1	2	3	4	5	6	7	8	9	10	11	12	13	14	15	16	17	18
2	4	6	8	10	12	14	16	18	20	22	24	26	28	30	32	34	36
3	6	9	12	15	18	21	24	27	30	33	36	39	42	45	48	51	54
4	8	12	16	20	24	28	32	36	40	44	48	52	56	60	64	68	72
5	10	15	20	25	30	35	40	45	50	55	60	65	70	75	80	85	90
6	12	18	24	30	36	42	48	54	60	66	72	78	84	90	96	102	108
7	14	21	28	35	42	49	56	63	70	77	84	91	98	105	112	119	126
8	16	24	32	40	48	56	64	72	80	88	96	104	112	120	128	136	144
9	18	27	36	45	54	63	72	81	90	99	108	117	126	135	144	153	162
10	20	30	40	50	60	70	80	90	100	110	120	130	140	150	160	170	180
11	22	33	44	55	66	77	88	99	110	121	132	143	154	165	176	187	198
12	24	36	48	60	72	84	96	108	120	132	144	156	168	180	192	204	216
13	26	39	52	65	78	91	104	117	130	143	156	169	182	195	208	221	234
14	28	42	56	70	84	98	112	126	140	154	168	182	196	210	224	238	252
15	30	45	60	75	90	105	120	135	150	165	180	195	210	225	240	255	270
16	32	48	64	80	96	112	128	144	160	176	192	208	224	240	256	272	288
17	34	51	68	85	102	119	136	153	170	187	204	221	238	255	272	289	306
18	36	54	72	90	108	126	144	162	180	198	216	234	252	270	288	306	324

Fig. 2.1 Ready reckoner

If you have trouble with your times tables make a copy of Figure 2.1 and keep it with you. If you use the ready reckoner on a regular basis it should help you to remember more and more of your times tables. Once you feel confident multiplying numbers you can discard it.

The ready reckoner is quite simple to use. For example, to multiply 18 by 12, place a rule or straight edge along the bottom of the line marked 18 and run your finger down the 12 column; the number found at the point where your finger meets the rule, 216, is the answer.

Exercise 2.1

Use your ready reckoner to find the answers to the following:

(a) 6×9
(b) 7×16
(c) 9×18
(d) 14×14
(e) 12×9
(f) 11×18
(g) 18×18
(h) 4×17
(i) 13×9
(j) 6×15
(k) 7×8

2.2 Electronic calculators

The ready reckoner on the previous page will, it is hoped, help you with the simpler calculations you encounter while working through this book, and at the same time encourage you to learn your multiplication tables. However, as mentioned in the introduction, it is highly recommended that as you work through the book you also have the use of an electronic calculator.

Calculators are manufactured in a wide and constantly changing variety of models, as a visit to your local stationers will show. Do not be tempted, when faced with what is often a large shop display, into buying a calculator that has many more functions than you require or has a keyboard that is too small; remember, a calculator will only compute the information that is entered into it and there can be a tendency to press the wrong keys on some of the smaller or multi-functioned calculators. With this in mind, unless you are certain, a rough check as outlined in Chapter 3 should always be made on the calculated answer given.

I have, before starting each example in this chapter, carried out a rough estimate on the size of the answer. **This is a strategy you should employ each time you use a calculator**. Do not worry if, at this stage, you do not quite understand how this is achieved, but do return to this chapter after reading Chapter 3.

Listed below, and briefly explained, are the minimum functions your calculator should contain.

[+]	Add key	Instructs the calculator to add the next entered value to the displayed number.
[−]	Subtract key	Instructs the calculator to subtract the next entered value from the displayed number.
[×]	Multiply key	Instructs the calculator to multiply the displayed number by the next entered value.
[÷]	Divide key	Instructs the calculator to divide the displayed number by the next entered value.
[√‾]	Square root key	Calculates the square root of the displayed number.
[x^2]	Square key	Calculates the square of the displayed number.
[π]	Pi key	Enters the value of pi (≈ 3.1415927). *Note*: Not all calculators have [x^2] and [π] keys.
[C]	Clear key	Clears the current calculation. Should be used before each new calculation is started.
[AC]		Usually turns the calculator on/off.
[M+]	Memory add key	Instructs the calculator to add displayed number to contents of memory.
[M−]	Memory minus key	Instructs the calculator to subtract displayed number from contents of memory.

Note: Do not worry if your calculator does not have the [M+] and [M−] keys. It simply means it conforms to algebraic rules.

[MI]	Memory in	Moves the displayed number into the memory.
[MR]	Memory recall key	Displays the contents of the memory.
[=]	Equals key	Instructs the calculator to carry out the mathematical process and display the answer.

And, of course, the ten number keys marked 0, 1, 2, 3, 4, 5, 6, 7, 8, 9 and a decimal point key. (*Note*: If your calculator consistently gives different answers from those given in this book, your calculator may use different logic and you should check the instruction manual.)

Let us now work through some examples. I have purposely kept to low numbers in the examples so that you can more easily follow the logic of what is happening.

Example

Find the value of

6.93 + 4.18 + 20.3 + 10.7

(rough estimate 7 + 4 + 20 + 10 = 41)

Keys	Display
[6] [.] [9] [3]	6.93
[+]	6.93
[4] [.] [1] [8]	4.18
[+]	11.11
[2] [0] [.] [3]	20.3
[+]	31.41
[1] [0] [.] [7]	10.7
[=]	42.11

Notice that each time the [+] key is pressed the calculator displays the total of the calculation at that stage. This can be particularly useful if you are keeping a running total.

Example

Find the value of

17.22 − 6.89 − 3.142

(rough estimate 17 − 7 − 3 = 7)

Keys	Display
[17] [.] [2] [2]	17.22
[−]	17.22
[6] [.] [8] [9]	6.89
[−]	10.33
[3] [.] [1] [4] [2]	3.142
[=]	7.188

Again each time the [−] key is pressed the calculator displays the total of the calculation at that stage.

Example

Find the value of

3.4 × 4.2 × 53.2

(rough estimate 3 × 4 × 53 = 636)

Keys	Display
[3] [.] [4]	3.4
[×]	3.4
[4] [.] [2]	4.2
[×]	14.28
[5] [3] [.] [2]	53.2
[=]	759.696

Again a running total is displayed.

Example

Find the value of

$$12.34 \div 3.5$$

(rough estimate $12 \div 4 = 3$)

Keys	Display
[1] [2] [.] [3] [4]	12.34
[÷]	12.34
[3] [.] [5]	3.5
[=]	3.5257143

The answer your calculator gives to this example will depend on the size of the display. Some calculators have a ten-figure display and will give the answer as 3.525714286. Both answers are correct for our purposes.

2.3 Standard algebraic rules and the calculator

In Chapter 1 we looked at the order of mathematical operations and said that multiplication and division have equal priority, as do addition and subtraction. Therefore, a calculation that contains only either addition and subtraction or multiplication and division can often be entered into the calculator as it is presented.

Examples

1 Find the value of

$$24.56 + 16.34 - 19.35$$

(rough estimate $(25 + 16) - 19 = 22$)

Keys	Display
[2] [4] [.] [5] [6]	24.56
[+]	24.56
[1] [6] [.] [3] [4]	16.34
[-]	40.9
[1] [9] [.] [3] [5]	19.35
[=]	21.55

2 Find the value of

$$\frac{8.6 \times 3.2}{4}$$

$$\left(\text{rough estimate } \frac{9 \times 3}{4} = 6.75\right)$$

Key	Display
[8] [.] [6]	8.6
[×]	8.6
[3] [.] [2]	3.2
[÷]	27.52
[4]	4
[=]	6.88

However, it was also said to follow the standard algebraic rules: i.e. the multiplication or division in a sum must be carried out before any addition or subtraction is attempted and the majority of calculators have this ability programmed into them as standard. Now would be a good time to find out if your calculator has this ability.

Example

Find the value of $3 + 2 \times 4$

Key	Display
[3] [+] [2] [×] [4] [=]	11

If your calculator gave the answer as 20, then it does not follow algebraic rules. This is not normally a problem, except that you will need to look carefully at how mixed calculations are entered into the calculator before starting, making sure that any multiplication or division is carried out before addition or subtraction. Better still, rewrite the example with brackets, as explained in Chapter 1, before entering into the calculator.

Example

Find the value of $(2 \times 4) + 3$

Key	Display
[2] [×] [4] [+] [3] [=]	11

It is not important which calculator you purchase or use as long as you are aware of how it functions.

Often the memory keys can be used to good advantage. When you store a value in memory, the letter 'm' appears in the display. If you store a zero in memory or if you subtract a value to the number already there that results in a value of zero, the letter 'm' disappears. When you store another value in memory, it replaces the current memory.

The next example shows the same calculation carried out with a calculator that follows algebraic rules and one that does not.

Example

Find the value of

$$\frac{6 \times 2}{4} + \frac{5 \times 10}{5}$$

$$\left(\text{rough estimate } \frac{6 \times 2}{4} + \frac{5 \times 10}{5} = 13 \right)$$

Algebraic logic		**No algebraic logic**	
Key	*Display*	*Key*	*Display*
[6] [×] [2]		[6] [×] [2]	
[÷]	12	[÷]	12
[4]	4	[4]	4
[+]	3	[M+]	m3
[5] [×] [10]		[5] [×] [10]	
[÷]	50	[÷]	m50
[5]	5	[5]	
[=]	13	[M+]	m10
		[MR]	m13

There would, of course, be nothing against working out the calculation bit by bit and jotting the answers, as they are obtained, down on paper and putting the results back into the calculator to find the final answer. However, the more times a number comes out of a calculator and is put back the more chances there are for mistakes. Remember also that, at the moment, we are only dealing with relatively small numbers, but as we progress through the book the numbers, and the opportunities to make mistakes, will become greater.

The square [x^2] and square root [$\sqrt{}$] keys are explained later in the book as they are used.

When using a calculator for the first time, time spent mastering its use and reading its accompanying booklet will be paid back many times and if, like me, you believe that mathematics touches our lives every day you should get full use from it. There are no self-assessment questions associated with this chapter; instead practise using and understand using your calculator.

Answers to exercises

2.1   (a) 54; (b) 112; (c) 162; (d) 182; (e) 108; (f) 198; (g) 324; (h) 68; (i) 117; (j) 90; (k) 56.

3.1 The decimal system of numbers
3.2 Addition and subtraction of decimals
3.3 Multiplication of decimals
3.4 Division of decimals
3.5 Decimal places
3.6 Rough estimates

Prior Knowledge If you feel you are already skilled in the subjects mentioned above, turn to page 27 and try working through the SAQs. If you find difficulty in answering them, work through this chapter.

3.1 The decimal system of numbers

In Chapter 2 we looked at numbers and learnt that counting involved the use of only ten symbols or digits 0, 9, 8, 7, 6, 5, 4, 3, 2, 1 and a system of place value. As it had a base of ten we called it a decimal system. It was said that in a number each digit has a place value, and the position of that digit in the number indicated its value. Starting from the right-hand side of the number the digits represented ones, tens, hundreds, thousands, tens of thousands, hundreds of thousands and so on. It may be easier, until you are more familiar with them, to think of these place values as columns. Each column has ten times the value of the column to its right and a number placed in the column would immediately show its value.

Example

The number 635,724 would be shown as

100,000	10,000	1,000	100	10	1
6	3	5	7	2	4

which shows that there are

6 lots of 100,000	or	600,000
3 lots of 10,000	or	30,000
5 lots of 1,000	or	5,000
7 lots of 100	or	700
2 lots of 10	or	20
4 lots of 1	or	4

Giving the number 'Six hundred and thirty-five thousand, seven hundred and twenty-four' (635,724).

This system of indicating the value of a digit, where its value depends on the place it occupies, can be extended in the same way to the right of the '1' column:

100,000	10,000	1,000	100	10	1	•	tenths	hundredths	thousandths

The new columns on the right show places for tenths, hundredths and thousandths. A decimal point is used to show which numbers indicate whole units and which numbers represent parts of a unit. The decimal point is a dot (•) and is placed immediately after the '1s' digit.

It may be easier to think of these columns as representing 1 divided into ten, one hundred and one thousand equal parts, or

$1 \div 10$, which gives us the decimal 0.1, which is a tenth of 1
$1 \div 100 = 0.01$ a hundredth of 1
$1 \div 1,1000 = 0.001$ a thousandth of 1

A number, say 8, placed in the tenths column would represent a value of eight parts of a 1 divided into ten equal parts or $8 \times 0.1 = 0.8$. An eight in the hundredths column would represent a value of eight hundredths of 1 or $8 \times 0.01 = 0.08$, and so on. These columns are decreasing by ten times the value of the column to the left. The tenths, hundredths and thousandths are best written in their decimal form as 0.1, 0.01 and 0.001.

If there are no whole numbers it is good practice to place a zero in front of the decimal point. This does not alter its value, it simply avoids confusion and confirms the presence and position of the decimal point.

Placing zeros to the right of a decimal number does not change its value, as shown below.

Example

The decimal numbers 0.8 and 0.80 represent the same value:

100,000	10,000	1,000	100	10	1	•	0.1	0.01	0.001
					0	•	**8**	**0**	

The second number 0.80 is just confirming what you already know, that there is nothing in the hundredths columns.

However, zeros within a number must always be recognised and written. In the number three thousand and eighty-four (3,084), the zero keeps the place for the missing hundreds. Without it the number would be reduced tenfold to three hundred and eighty-four.

Example

The number 003809.370 could be represented as

100,000	10,000	1,000	100	10	1	•	0.1	0.01	0.001
0	0	3	8	0	9	•	3	7	0

which shows that there are

0 lots of 10,000
3 lots of 1,000
8 lots of 100
0 lots of 10
9 lots of 1
3 lots of 0.1
7 lots of 0.01
0 lots of 0.001

Giving the number 'Three thousand, eight hundred and nine point three seven'.

Again the zeros at either end of the number are surplus, but the zero in the tens column not only tells us there are no tens, but also keeps the other numbers in their correct place. The number would be better written as 3809.37.

3.2   Addition and subtraction of decimals

The addition and subtraction of decimal numbers is carried out in the same way as for whole numbers. The only thing to remember is that when writing out the calculation and the answer, care must be taken to keep the decimal points directly underneath each other as I have done in the next four examples. This will ensure that all the place values, the tenths, hundredths, etc., are lined up in the same vertical column.

Examples

1   10.37 + 405.31 + 0.423 would be written as

```
    10.37
   405.31
+    0.423
   416.103   Notice the decimal points all line up with each other.
```

2   31.43 + 2.218 would be written as

```
    31.43
 +   2.218
    33.648
```

3   4.964 − 2.531 would be written as

```
    4.964
 −  2.531
    2.433
```

4   6.731 − 2.62 would be written as

$$
\begin{array}{r}
6.731 \\
-\ 2.620 \\
\hline
4.111
\end{array}
$$

Notice that, in the last example, I placed a zero at the end of the number 2.62. This can sometimes prove useful as it keeps the numbers and the decimal point in their proper place, as was shown. Remember it does not alter the value of the number.

 **Exercise 3.1**

Write down and complete:

(a) $3.302 + 0.07 + 2.101$
(b) $123.4 + 12.34 + 1.234$
(c) $27.001 − 21.01$
(d) $2.357 − 0.292$

3.3 Multiplication of decimals

When multiplying numbers that contain decimals, the decimal points are either removed or ignored, leaving just whole numbers to be multiplied. Once the multiplication has been carried out and an answer obtained, a note is made of the number of figures following the decimal points in the original two numbers. The position of the decimal point in the answer is found by counting this number of figures from the right.

Example

1 3.52×1.3

$$
\begin{array}{r}
352 \\
\times\ \ 13 \\
\hline
1056 \\
352 \\
\hline
4576
\end{array}
$$
Ignoring the decimal points multiply the numbers together.

$$
\begin{array}{l}
3.52 \\
1.3 \\
2 + 1 = 3
\end{array}
$$
Add the total number of figures following the decimal point in both numbers: 3.52 has two numbers after the decimal point (i.e. 5 and 2) and 1.3 has one number after the decimal point (i.e. 3).

$4._3 5_2 7_1 6$ Count three places from the right and insert the decimal point.

4.576

2 1.374 × 0.23

$$
\begin{array}{r}
1374 \\
\times\ \ \ 23 \\
\hline
4122 \\
27480 \\
\hline
31602 \\
\end{array}
$$

Multiply the two numbers together, ignoring the decimal (points. Notice that I have left the zero out of the second number, as its only function is to confirm the absence of any whole numbers.)

1.374
0.23
$3 + 2 = 5$

Add up the total number of figures after the decimal points in both numbers to be multiplied: 1.374 has three numbers after the decimal place (3, 7, 4) and 0.23 has two (2, 3), making a total of five.

$._5 3_4 1_3 6_2 0_1 2$

Count five figures from the right of the answer and insert the decimal point.

0.31602

I have put a zero in front of the decimal point as there are no whole numbers.

3.4 Division of decimals

If the divisor (the number you are dividing by) is a decimal, it must first be changed into a whole number by removing the decimal point. This can be achieved by multiplying the divisor by 10 as often as necessary. Each time the multiplication is carried out, the decimal point will move one place to the right, and this is continued until the decimal point is on the far right of the number. The number to be divided must also be multiplied by 10 *the same number of times* as was the divisor, adding zeros if necessary.

Example

3.3 ÷ 0.15

$0.15 \times 10 = 1.5$ I had to multiply by 10 twice to turn 0.15 into a whole number.
$1.5 \times 10 = 15$

I will need to do the same to 3.3:

$3.3 \times 10 = 33$
$33 \times 10 = 330$

The calculation now becomes a simple division:

330 ÷ 15

$$
\begin{array}{r}
022 \\
15\overline{)330} \\
30 \\
\hline
30 \\
\end{array}
$$

Therefore, 3.3 ÷ 0.15 = 22

Really all we did to change 0.15 into a whole number was to move the divisor two places to the left (multiplying by 10 is the same as moving the divisor one place to the left) and as we increased its value from 0.15 to 15 the same had to be done to 3.3 by moving it two places to the left, thereby changing it to 330.

Rather than multiply by 10 each time, count the number of times the divisor is required to move to obtain a whole number and move the number to be divided the same number of times. (*NB:* The moving of numbers about the decimal point is explained fully in Chapter 6.)

Example

$1.464 \div 1.2$

The divisor needs to move once to the left to change 1.2 into the whole number 12. So the number 1.464 is also moved once to the left, to give 14.64 and the calculation now becomes $14.64 \div 12$.

Care must be taken when writing down the calculation and its answer, as the position of the decimal point in the answer must be *directly above* its position in the number to be divided.

```
    01.22     As you cannot divide 1 by 12, a zero is placed in the answer above the 1
12 )14.64     14 ÷ 12 = 1 (place a 1 in the answer)
    12
     2        with 2 over
    26        bring down the 6: 26 ÷ 12 = 2 (place a 2 in the answer)
     2        with 2 over
    24        bring down the 4: 24 ÷ 12 = 2 (place a 2 in the answer)
```

The zero was placed in the answer while working out the example to help keep the figures and decimal points in line. As it has no value or direct bearing on the answer, it can now be removed, leaving 1.22.

The above example worked out to an exact answer, i.e. it had no remainder, but this is not always the case. Many divisions, when worked out, seem to have no end to them and in fact some do not (try dividing 22 by 7).

Sometimes before you work out a division calculation it is worth thinking about how accurate your answer needs to be. After all, when working on a building site most trades would consider working to plus or minus a millimetre to be very accurate and measuring anything smaller than this with a tape is almost impossible.

Let's do another example, but this time think about how accurate the answer needs to be.

Example

$46.836 \div 2.41$

First move the number two places to the left, to convert 2.41 into a whole number:

```
        0019.434024
241 | 4683.6              241 will not divide into either 4 or 46, so two zeros are placed in
      241                 the answer.
      227                 468 ÷ 241 = 1 with 227 over.
      2273                Bring down the 3 to give 2273 ÷ 241 = 9 with 104 over
      2169                (241 × 9 = 2169).
       104
      1046                Bring down the 6 to give 1046 ÷ 241 = 4 with 82 over
       964                (241 × 4 = 964).
        82
       820                As all the figures have been used, bring down a 0 to give
       723                820 ÷ 241 = 3 with 97 over (241 × 3 = 723).
        97
       970                Bring down a 0 to give 970 ÷ 241 = 4 with 6 over
       964                (241 × 3 = 964).
         6
        60                Bring down a 0 to give 60; but 241 will not go into 60 so a 0 is
                          placed in the answer.
       600                Another 0 is brought down to give 600 ÷ 241 = 2 with 118 over
       482                (241 × 2 = 482).
       118
      1180                Bring down a 0 to give 1180 ÷ 241 = 4 with 216 over.
       964
       216
```

The answer to six decimal places is 19.434024. This cannot be called the correct answer as there is a remainder (216). We can say, however, that

$$46.836 \div 2.41 \approx 19.434024$$

(Remember that the sign means approximately equal to.)

3.5 Decimal places

The division example could have been continued to give as many decimal places as desired, or until there was no remainder, but I decided that six decimal places was as accurate as the answer needed to be – that is, there are six figures after the decimal point. If we now decide that we only require our answer to be correct to three decimal places, the number 19.434027 would become 19.434.

When limiting a number to a certain number of places after the decimal point, the usual rule is: *if the first figure to be discarded is less than five it is ignored; if it is five or larger; then the previous figure is increased by one*. For example, if the number was 19.434627 it would become 19.435, as the 6 is larger than 5, one is added to the previous figure, 4. It is now correct to three decimal places.

If an answer is required to three decimal places, you must continue the division to *four* decimal places and correct the answer back to three.

Exercise 3.2

Rewrite the following, correct to three places of decimal:

(a)   135.3646
(b)   7.473968
(c)   3.98531
(d)   1.2004

It can sometimes be a problem deciding how many decimal places to work to or how accurate the answer should be. This is often a case of exercising judgement when giving your answer. Let's look at the number 19.434024 and think of it as the proposed length of a brick-built wall.

The first two figures, 19, indicate units as they are on the left of the decimal point. The unit of measurement we shall use is the metre (the metre is the basic SI unit of length measurement; it is divided into a thousand millimetres. Although we shall be learning more about it in Chapter 4, now will be a good time (if you don't already know) to familiarise yourself, with just how big or small these measurements are.)

The 4 just after the decimal point indicates four tenths of a unit, or in our case four tenths of a metre.

A tenth of a metre is 1,000 millimetres ÷ 10 = 100 millimetres.
Therefore, the 4 represents 4 × 100 = 400 millimetres.
This certainly seems to be big enough to take into account.

The second figure after the decimal point, 3, is in the hundredths position and tells us there are three hundredths of a metre.

A hundredth of a metre is 1,000 millimetres ÷ 100 = 10 millimetres.
Therefore, the 3 represents 3 × 10 = 30 millimetres.
This still seems big enough to take into account.

The third figure after the decimal point, 4, is in the thousandths position and indicates that there are four thousandths of a metre.

A thousandth of a metre is 1,000 millimetres ÷ 1000 = 1 millimetre.
Therefore the 4 represents 4 × 1 = 4 millimetres.
This is just about big enough to take into account with the brick wall.

The fourth and fifth figures represent ten thousandths and hundred thousandths of a metre, or a tenth and a hundredth of a millimetre. I am sure you will agree that we would have difficulty measuring the wall to that level of accuracy. Certainly for all practical purposes we can forget about the figures after three places past the decimal point since each one represents a length which is one tenth of the previous one. At the third decimal place we are dealing with lengths little more than the thickness of a pencil line.

This leaves us with the measurement 19.434 metres or nineteen metres and four hundred and thirty-four millimetres.

 Exercise 3.3

Give all your answers (where applicable) to three places of decimal.

(a) 68.23 × 200
(b) 346.214 × 1.7
(c) 37.149 × 2.74
(d) 2.1 × 0.375
(e) 121 ÷ 5.5
(f) 23.184 ÷ 3.6
(g) 6.395 ÷ 2.7
(h) 0.22 ÷ 0.7

3.6 Rough estimates

It is considered good practice to make a rough estimate of your answer before the calculation is started. This can help to avoid one of the worst mistakes often made when working with decimal points – that is, to place the point in the wrong position. Remember that a decimal point wrong by one place can make an answer ten times too large or ten times too small, which is disastrous if you are ordering materials or pricing work. One method of carrying out this is to correct the numbers to no decimal places.

Example

If you were required to multiply 12.37 by 4.81, the 12.37 would become 12 and the 4.81 correct to no decimal places would become 5: 5 × 12 = 60. So the answer should be approximately 60.

```
    1237
 ×   481
    1237
   98960
  494800
  594997
```

Counting four decimal places from the right gives the answer

    59.4997

The rough answer was 60, therefore the decimal point in the answer must be in the correct position and the answer is 59.4997 and not 5.94997 or 594.997.

Another example is given here.

Example

12.37 ÷ 4.81 would become 12 ÷ 5 = 2 approximately.

12.37 ÷ 4.81 = 2.5717256

Again we can see the decimal point in the answer is in the correct position.

 Exercise 3.4

Make a rough estimate of the following before finding the answers correct to three decimal places:

(a) 3.17×4.98
(b) 24.8×1.8
(c) 2.59×1.23
(d) $184.24 \div 4.29$
(e) $276.49 \div 5.96$
(f) $8.7 \div 3.2$

Before we leave this chapter on decimals lets just quickly look at a subject close to most builders' hearts: money. We have in Britain a decimal system of currency that uses the pound (£) as its basic unit. This basic unit is broken down further into a sub-unit called the penny (p) with 100 pence being equal to one pound. A decimal point is used to separate the basic units from the sub-units, for example £7.23 means seven pounds and twenty-three pence.

As it is a decimal system, all that you have just learnt about addition, subtraction, multiplication and division apply when working with money, as the following examples show.

A painter buys three cans of paint. What is the total cost?

1 can of emulsion costs	£14.78	Notice that the decimal points and
1 can of undercoat costs	£5.79	the units have been placed beneath
1 can of gloss paint costs	£8.06	each other.
	£28.63	

That is twenty-eight pounds and sixty-three pence.

The customer had given the painter £35.00 to buy the paint and now requires the change.

Money forwarded to painter	£35.00	Again, the decimal points and
Cost of paint	£28.63	the units have been placed
Change required	£6.37	directly beneath each other.

The painter also has to buy 14 rolls of wallpaper at £10.50 a roll. What will be the cost of this?

1 roll of paper	£10.50	I have left the sign and decimal point
Total rolls required	14	to show it is the money we are dealing
	1050	with.
	4200	
	£147.00	There are two places past the decimal point, i.e. 50.

Note: A rough estimate would have given 11 × 14 = 154.

The painter arrives at the merchant only to find the total cost of the paper has increased to 156.80. What now will be the cost of a single roll of paper. To find this we need to divide the total cost of the paper by the number of rolls brought.

```
        11.20
    14│156.80
        14          15 ÷ 14 = 1 with 1 over
        ──
         1
        16          bring down the 6: 16 ÷ 14 = 1
        14
        ──
         2
        28          bring down the 8: 28 ÷ 14 = 2
        28
        ──
        00
        00          bring down the 0: 0 ÷ 14 = 0
```

The paper costs £11.20 a roll.

Exercise 3.5

(a) Add together £32.45, £19.46 and £12.14
(b) Add together £10.49, £91.04 and £0.62
(c) Subtract £9.34 from the following: £28.12, £104.23 and £11.05
(d) Find the total cost of 19 lengths of copper tube if each one costs £3.27
(e) If 16 sheets of plasterboard cost £92.80, What would be the cost of one sheet?

Answers to exercises

3.1

(a)	3.302	(b)	123.4	(c)	27.001	(d)	2.357
	0.070		12.34		−21.01		−0.292
	2.101		1.234		5.991		2.065
	5.473		136.974				

3.2   (a) 135.365; (b) 7.474; (c) 3.985; (d) 1.200

3.3   (a) 13646; (b) 588.564; (c) 101.788; (d) 0.788 (e) 22; (f) 6.44; (g) 2.369; (h) 0.314

3.4

Rough check	*Correct answer*
(a) $3 \times 5 = 15$	15.787
(b) $25 \times 2 = 50$	44.64
(c) $3 \times 1 = 3$	3.186
(d) $184 \div 4 = 46$	42.946
(e) $276 \div 6 = 46$	46.391
(f) $9 \div 3 = 3$	2.719

3.5   (a) £64.05; (b) £102.15; (c) £18.78, £94.89, £1.71 (d) £62.13; (e) £5.80

Summary

We have been looking at:

- The decimal system of numbers.
- The addition and subtraction of decimal numbers.
- The multiplication and division of decimal numbers.
- Rough checks.

Things to remember

Addition of decimals

- Take care to arrange the decimal points directly below each other in the numbers to be added.
- Carry out the addition as you would with whole numbers, putting the decimal point in the answer directly below the decimal points in the calculation.
- This ensures that all the place values, the tenths, hundredths, etc., are lined up vertically.

Subtraction of decimals

- Take care to arrange the decimal points directly below each other in the numbers to be subtracted.
- Carry out the subtraction as you would with whole numbers, putting the decimal point in the answer directly below the decimal points in the calculation.
- This ensures that all the place values, the tenths, hundredths, etc., are lined up vertically.

Multiplication of decimals

- Disregard the decimal points in the numbers and multiply as you would with whole numbers.
- The number of decimal places in the answer is equal to the sum of the number of decimal places in the two numbers to be multiplied.

Division of decimals

- If the divisor is not a whole number, multiply it by 10 or move it to the left as often as necessary until it becomes a whole number.
- Whether the number to be divided is a whole number or not, it should be moved to the left, or multiplied by 10, the same number of times as the divisor.
- Divide as you would for whole numbers, placing the decimal point in the answer directly above the decimal point in the number to be divided.

Decimal places

Decide how accurate or how many decimal places you need to work to, then apply the rule: if the first figure to be discarded is less than 5 it is ignored; if it is 5 or larger then the previous figure is increased by one.

Rough estimates

Correct the numbers to no decimal places and carry out the required calculation. Use the answer as a guide for the position of the decimal place in the correct answer.

Take a break from your studies before testing your skills on the following self-assessment questions.

Self-assessment questions

Below there are five questions. Take your time in answering them.

- These are not meant as a test; the questions are simply to help you learn.
- Look back at you own notes and Chapter 3 if you need help.
- Answers and comments follow the questions, but you should look at them only when you have finished or are really stuck.

Try working out the answers without the use of a calculator.

SAQ 3.1   Give the answers to:

(a) $45.0053 + 21.549$
(b) $902.34 + 0.3325$

(c) 12.57 − 6.36
(d) 3.945 − 0.16

SAQ 3.2   Give the answers (where applicable, correct to two decimal places) to:

(a) 28.9 ÷ 9
(b) 8.46 ÷ 2.2

Make rough estimates of the next two questions before attempting them:

(c) 354.2 × 3.6
(d) 236 × 2.4

SAQ 3.3   Below is a list of prices for materials a carpenter requires to complete a job. What is the total cost of materials?

1 door at £38.20
1 set of handles at £4.30
1 mortise lock at £9.45
1 pair of hinges at £5.68

SAQ 3.4   A bonus of £791.56 is to be shared equally among seven bricklayers. How much will each bricklayer receive?

SAQ 3.5   The bricklayers decide to give £30 each to a fund to be shared out to the five hod carriers working with them. How much money will be in the fund and how much will each hod carrier receive?

Answers and comments

SAQ 3.1

(a) $\begin{array}{r} 45.0053 \\ + 21.5490 \\ \hline 66.5543 \end{array}$ The zero has been placed after the 9 simply to help keep the numbers and decimal point in their proper places.

(b) $\begin{array}{r} 902.34 \\ + \ \ 0.3325 \\ \hline 902.6725 \end{array}$ Notice how I have kept the decimal points directly underneath each other. Making sure that all the place values, the tenths, hundredths, etc., are lined up vertically.

(c) $\begin{array}{r} 12.57 \\ - \ \ 6.36 \\ \hline 6.21 \end{array}$ Again the decimal points must be directly underneath each other.

(d) $\begin{array}{r} 3.945 \\ - 0.16 \\ \hline 3.785 \end{array}$

SAQ 3.2

(a)
```
       3.211
    9│28.9
      27
       19      Bring down the 9.
       18
        10     Bring down a 0.
         9
         10    Bring down a 0.
```

The question asks for the answer correct to two decimal places. I have continued the division to three decimal places and can now correct it to two, to give the answer 3.21.

(b)
```
        03.845
     22│84.6        The numbers have been moved one place to the left, to
        66          change 2.2 into a whole number.
        186         Bring down the 6.
        176
         100        Bring down a 0.
          88
         120        Bring down a 0.
```

Answer: 3.85 correct to two decimal places.

(c) Rough estimate: $354 \times 4 = 1416$

```
     3542      I have left out the decimal points, but have noted that there
   ×   36      are two figures in total after the decimal points in the
   21252       question.
   10626
  127512       Count two places in from the right, insert the decimal point,
               to give the answer 1275.12.
```

The rough estimate confirmed that the decimal point is in the correct place.

(d) Rough estimate: $236 \times 2 = 472$

```
     236       I have left out the decimal points, but have noted that there
   ×  24       is one figure after the decimal points in the question.
     944
     472
    5664       Count one place in from the right, insert the decimal point,
               to give 566.4.
```

The rough estimate confirmed that the decimal point is in the correct place.

SAQ 3.3

```
  £38.20
   £4.30        Make sure the decimal points are lined up.
   £9.45
   £5.68
  £57.63
```

SAQ 3.4

To find how much bonus each bricklayer receives, the 791.56 is divided by 7.

```
       113.08
    7 | 791.56
       7
       09          Bring down the 9.
        7
       21          Bring down the 1.
       21
        05         Bring down the 5. 7 ÷ 5 will not go, add a 0 to the answer.
        56         Bring down the 6.
```

Each bricklayer will receive £113.08 bonus.

SAQ 3.5

There are seven bricklayers and they each give £30 to the fund, making a total of £30 × 7 = £210 in the fund. By dividing the fund (£210) by the number of hod carriers (5) the amount of each hod carrier receives can be found:

```
       42
    5 | 210
      20
      10         Bring down the 0.
```

The hod carriers each receive £42.

CHAPTER 4 Fractions

Prior knowledge If you feel you are already skilled in the subjects mentioned above, turn to page 41 and try working through the SAQs. If you find difficulty in answering them, work through this chapter.

4.1 Fractions

As you may have noticed in the previous chapters, the decimal system of numbers, especially when using a calculator, is relatively straightforward. As we progress through the book you will find it is a number system that lends itself to the building trade, particularly when we start learning about and working with volumes and areas. However, using, as this book does, principally the decimal system of numbers does not mean we should or can dispense entirely with fractions. It is important that you understand the meaning of fractions, feel confident in their manipulation and be able to convert them to their decimal form.

Fractions should not be thought of as belonging exclusively to the imperial system of measurement of feet and inches, but rather as another method of expressing parts of a whole in the same way as the numbers following the decimal place.

If you broke a bar of chocolate into two pieces of the same size and gave one piece to a friend, he or she would have half of the chocolate. This could be written as a fraction, $\frac{1}{2}$. The lower number (2) is called the **denominator** and is saying that the chocolate bar is in two parts, while the top number (1), known as the **numerator**, indicates the number of parts of the chocolate bar that have been given away. Below, we shall look at two types of fractions: proper fractions and improper fractions.

4.2 Proper fractions

If a length of copper tube is cut into eight equal parts, each part would be $\frac{1}{8}$

of the original length. Take away five of these parts and you would have taken away $\frac{5}{8}$ of the original length, with $\frac{3}{8}$ remaining.

If, as above, the numerator is smaller than the denominator, then the fraction is called a **proper fraction**.

Some examples of proper fractions are $\frac{3}{8}, \frac{7}{16}, \frac{9}{16}, \frac{11}{32}$, all of which are less than one. It is this fact that enables them to be called *proper* fractions.

Fractions should always, if possible, be cancelled down to their lowest term. This can be achieved by dividing both the numerator and the denominator of the fraction by a factor that is common to both, preferably their *highest* common factor, known as their HCF. For example, the fraction $\frac{9}{24}$ can be reduced to its lowest term by dividing both the 9 and the 24 by 3 (their HCF) to give the fraction $\frac{3}{8}$.

Fractions expressed in their lowest terms are generally easier to work with. The fractions $\frac{3}{8}$ and $\frac{9}{24}$ both still represent the same value.

4.3 Improper fractions

If we can have proper fractions, it follows that we should also have **improper fractions**. These occur when the numerator is *larger* than the denominator. Here are some examples of improper fractions: $\frac{3}{2}, \frac{7}{4}, \frac{20}{15}$. As all of these fractions are greater than one, they are called *improper*.

Improper fractions can be turned into mixed numbers – that is, whole numbers and proper fractions – by dividing the denominator into the numerator and putting the remainder over the denominator. For example

$$\frac{3}{2} = 1 + \frac{1}{2} = 1\frac{1}{2} \qquad \frac{7}{4} = 1 + \frac{3}{4} = 1\frac{3}{4} \qquad \frac{20}{15} = 1 + \frac{5}{15} = 1\frac{5}{15}$$

The fraction $1\frac{5}{15}$ can be reduced further by dividing the numerator and the denominator by the highest factor that is common to them both (their HCF), in this case 5, giving the answer $1\frac{1}{3}$. If there is no obvious factor to divide by, this can be done gradually using a small number common to them both, for example

$$\frac{72}{112} = \frac{72 \div 2}{112 \div 2} = \frac{36 \div 2}{56 \div 2} = \frac{18 \div 2}{28 \div 2} = \frac{9}{14}$$

When the denominator and numerator can no longer be divided by a number that is common to both of them, the fraction is said to be reduced to its lowest term.

4.4 Adding and subtracting fractions

There are five steps to remember when adding or subtracting fractions. They have been written out below and together we can work through several examples.

(a) Add together or subtract from each other all the whole numbers if the fractions are mixed.

(b) Find the lowest common denominator (LCD) for all the fractions. For example, the fractions $\frac{1}{2}$, $\frac{2}{3}$ and $\frac{1}{4}$ have 12 as the lowest common denominator of 2, 3 and 4, that is 12 is the lowest number that 2, 3 and 4 will divide into. (*Note*: Strictly speaking, this does not have to be the lowest factor, but the calculation will be made much easier if time is spent finding it. If no LCD can easily be found, then multiplying the denominators together will give a common denominator.)

(c) Divide the LCD of the fractions by the denominator of each of the fractions and multiply the result by the numerator. Put the answer over the LCD, and do this for each of the fractions. For example, the fractions $\frac{1}{2}$, $\frac{2}{3}$ and $\frac{1}{4}$, using 12 as the LCD, would become

$$\frac{1}{2} = \frac{(12 \div 2 = 6) \times 1}{12} = \frac{6}{12}$$

$$\frac{2}{3} = \frac{(12 \div 3 = 4) \times 2}{12} = \frac{8}{12}$$

$$\frac{1}{4} = \frac{(12 \div 4 = 3) \times 1}{12} = \frac{3}{12}$$

The fractions $\frac{6}{12}$, $\frac{8}{12}$ and $\frac{3}{12}$ have the same value as $\frac{1}{2}$, $\frac{2}{3}$ and $\frac{1}{4}$ but now have a common denominator.

(d) Add or subtract all the numerators.

(e) Add the whole numbers from step (a) and reduce the fraction to its lowest term.

Examples

1 $1\frac{2}{3} + 2\frac{3}{4}$

$1 + 2 = 3$	(a)	Add together all the whole numbers.
$\frac{2}{3} + \frac{3}{4}$	(b)	12 is the lowest common denominator of 3 and 4.
$12 \div 3 = 4 \times 2 = 8 = \frac{8}{12}$	(c)	Convert the fractions to a common denominator.
$12 \div 4 = 3 \times 3 = 9 = \frac{9}{12}$		
$\frac{8}{12} + \frac{9}{12} = \frac{17}{12}$	(d)	Add the numerators.
$3 + \frac{17}{12} = 3 + 1\frac{5}{12}$	(e)	Add the whole number from step (a) and reduce the fraction to its lowest term.

$$= 4\frac{5}{12}$$

2 $3\frac{3}{4} + 2\frac{1}{2} + 3\frac{5}{8}$

$3 + 2 + 3 = 8$	(a)	Add the whole numbers.
$\frac{3}{4} + \frac{1}{2} + \frac{5}{8}$	(b)	8 is the LCD of 4, 2 and 8.
$\frac{6}{8} + \frac{4}{8} + \frac{5}{8}$	(c)	Convert the fractions to a common denominator.
$\frac{15}{8}$	(d)	Add the numerators.
$8 + 1\frac{7}{8}$	(e)	Add the whole numbers from step (a) and reduce the fraction to its lowest term.

$$= 9\frac{7}{8}$$

Exercise 4.1

Find the answer to:

(a) $3\frac{1}{8} + 2\frac{1}{4}$

(b) $1\frac{5}{9} + 1\frac{1}{2}$

(c) $6\frac{4}{5} + 3\frac{1}{3}$

(d) $1\frac{1}{8} + 3\frac{1}{4} + 4\frac{1}{4}$

The same procedure is used when subtracting fractions.

Examples

1 $3\frac{5}{8} - 1\frac{2}{5}$

$3 - 1 = 2$	(a)	Subtract 1 from 3.
$\frac{5}{8} - \frac{2}{5}$	(b)	40 is the LCD of 8 and 5.
$\frac{24}{40} - \frac{16}{40}$	(c)	Convert the fractions to a common denominator.
$\frac{9}{40}$	(d)	Subtract the numerators.
$2 + \frac{9}{40}$	(e)	Add the number from step (a) and reduce the fraction to its lowest term (in this case the fraction is already in its lowest term).
$= 2\frac{9}{40}$		

2 $5\frac{3}{8} - 2\frac{1}{4} + 3\frac{7}{12}$

$5 - 2 + 3 = 6$	(a)	Add and subtract whole numbers. (*Note*: The whole numbers must be added or subtracted in the order that they appear in the sum; that is, 2 is subtracted from 5 and 3 is added to the result.)
$\frac{3}{8} - \frac{2}{4} + \frac{7}{12}$	(b)	24 is the LCD of 8, 2 and 12. (*Note*: keep the signs in the same position as in the original sum.)
$\frac{9}{24} - \frac{6}{24} + \frac{14}{24}$	(c)	Convert the fractions to a common denominator.
$\frac{17}{24}$	(d)	Add and subtract the numerators.
$6 + \frac{17}{24}$	(e)	Add the number from step (a) and reduce the fraction to its lowest term (once again the fraction is already in its lowest term).
$= 6\frac{17}{24}$		

Exercise 4.2

Find the answers to:

(a) $3\frac{9}{16} - 2\frac{3}{8}$

(b) $3\frac{5}{9} - 2\frac{7}{9}$

(c) $6\frac{1}{2} - 4\frac{3}{4} - 1\frac{1}{8}$

(d) $3\frac{1}{4} - 2\frac{13}{15}$

4.5 Multiplying fractions

Again there is a set of rules to remember. They have been written out below and we can work through examples together, step by step.

(a) All mixed numbers must be changed into improper fractions, by multiplying the whole number by the denominator, adding the numerator and putting the result over the original denominator. For example, to change $3\frac{1}{2}$ into an improper fraction

$$(3 \times 2 = 6) + 1 = 7 = \tfrac{7}{2}$$

and again to change $7\frac{9}{10}$ into an improper fraction

$$(7 \times 10 = 70) + 9 = 79 = \tfrac{79}{10}$$

(b) The calculation can be made much easier if the numerator and the denominator across from each other have a common factor. This can then be used to divide both the numerator and the denominator to their lowest terms. This is called cross cancelling.

(c) Multiply all the numerators together, and all the denominators together.

(d) If the result is an improper fraction, change it to a mixed number.

Examples

1 $3\frac{1}{3} \times 2\frac{1}{4}$

$(3 \times 3 = 9) + 1 = 10 = \tfrac{10}{3}$	(a) Change to improper fractions.
$(2 \times 4 = 8) + 1 = 9 = \tfrac{9}{4}$	
$\tfrac{10}{3} \times \tfrac{9}{4} = \tfrac{5}{1} \times \tfrac{3}{2}$	(b) The 3 and 9 will divide by 3 to give 1 and 3. The 10 and 4 will divide by 2 to give 5 and 2.
$\tfrac{5}{1} \times \tfrac{3}{2} = \tfrac{5 \times 3}{1 \times 2}$	(c) Multiply all the numerators together, and all the denominators together.
$\tfrac{15}{2}$	(d) As this is an improper fraction the numerator must be divided by the denominator to arrive at either a whole number or, as in this case, a mixed number.
$= 7\tfrac{1}{2}$	

2 $1\frac{1}{3} \times 2\frac{2}{3}$

$\tfrac{4}{3} \times \tfrac{8}{3}$	(a) Change to improper fractions.
	(b) As nothing will cross cancel out, step (b) in this example is ignored.
$\tfrac{4}{3} \times \tfrac{8}{3} = \tfrac{4 \times 8}{3 \times 3}$	(c) Multiply the numerators together and the denominators together.
$\tfrac{32}{9}$	(d) Change to a mixed number.
$= 3\tfrac{5}{9}$	

3 $1\frac{1}{2} \times 2\frac{1}{2} \times 3\frac{1}{3}$

$\frac{3}{2} \times \frac{5}{2} \times \frac{10}{3}$ (a)   Change to improper fractions.

$\frac{3}{2} \times \frac{5}{2} \times \frac{10}{3} = \frac{1}{1} \times \frac{5}{2} \times \frac{5}{1}$ (b)   One 2 and the 10 will divide by 2 leaving 1 and 5. The 3s cancel each other out.

$\frac{1}{1} \times \frac{5}{2} \times \frac{5}{1} = \frac{1 \times 5 \times 5}{1 \times 2 \times 1}$ (c)   Multiply the numerators together and the denominators together.

$\frac{25}{2}$ (d)   change to a mixed number.

$= 12\frac{1}{2}$

4 I have left the steps out in this example for you to fill in yourself and make your own notes.

$2\frac{1}{10} \times \frac{5}{7}$

$\frac{21}{10} \times \frac{5}{7}$

$\frac{21}{10} \times \frac{5}{7} =$

$\frac{3}{2} \times \frac{1}{1} = \frac{3 \times 1}{2 \times 1}$

$\frac{3}{2}$

$= 1\frac{1}{2}$

 Exercise 4.3

What is the answer to:

(a) $3\frac{1}{2} \times 1\frac{1}{2}$

(b) $2\frac{1}{4} \times 1\frac{1}{4}$

(c) $3\frac{1}{2} \times 1\frac{3}{16}$

(d) $1\frac{1}{2} \times 1\frac{1}{4} \times 1\frac{3}{4}$

4.6   Dividing fractions

Except for step (b), the steps to be taken for dividing fractions are similar to those for multiplying fractions. I have set them out below and will work through examples with you.

(a)   All mixed numbers must be turned into improper fractions.
(b)   The fraction you are dividing by is turned upside down and the division sign is changed to a multiplication sign.
(c)   Cross cancel if possible.
(d)   Multiply the numerators together and the denominators together.
(e)   If the result is an improper fraction, change to a mixed number.

Examples

1 $6\frac{3}{4} \div 1\frac{1}{8}$

$\frac{27}{4} \div \frac{9}{8}$ (a)   Change to improper fractions.

$\frac{27}{4} \times \frac{8}{9}$ (b)   Turn the divisor upside down, change the division sign to a multiplication sign.

$\frac{3}{1} \times \frac{2}{1}$ (c)   The 27 and the 9 have cancelled out, and the 4 and the 8 can be cancelled.

$\frac{3}{1} \times \frac{2}{1} = \frac{6}{1}$ (d)   Multiply the numerators together and the denominators together.

6   (e)   Change to mixed number or, in this case, a whole number.

2 I have left the note out in the last three examples. Write them in yourself in your own words.

$3\frac{1}{2} \div 1\frac{1}{4}$

$\frac{7}{2} \div \frac{5}{4}$ (a)

$\frac{7}{2} \times \frac{4}{5}$ (b)

$\frac{7}{1} \times \frac{2}{5}$ (c)

$\frac{14}{5}$ (d)

$2\frac{4}{5}$ (e)

3 $3\frac{3}{16} \div 1\frac{1}{4}$

$\frac{51}{16} \div \frac{5}{4}$ (a)

$\frac{51}{16} \times \frac{4}{5}$ (b)

$\frac{51}{4} \times \frac{1}{5}$ (c)

$\frac{51}{20}$ (d)

$2\frac{11}{20}$ (e)

4 $2\frac{1}{8} \div \frac{7}{16}$

$\frac{17}{8} \div \frac{7}{16}$ (a)

$\frac{17}{8} \times \frac{16}{7}$ (b)

$\frac{17}{1} \times \frac{2}{7}$ (c)

$\frac{34}{7}$ (d)

$4\frac{6}{7}$ (e)

Exercise 4.4

Find the answer to:

(a) $3\frac{1}{2} \div 1\frac{1}{2}$

(b) $2\frac{1}{4} \div 1\frac{1}{4}$

(c) $3\frac{1}{2} \div 1\frac{1}{16}$

(d) $1\frac{1}{2} \div 1\frac{1}{4}$

| 4.7 | Fractions to decimals |

Before starting the conversion of numbers expressed as fractions to decimals, it is always worthwhile considering the degree of accuracy or the number of decimal places the answer requires. Some answers to conversions can stretch into infinity, that is they have no end.

We shall work, as is common throughout this book, to three decimal places of accuracy. If you can, set your calculator to three decimal places. If not, remember from Chapter 2 that if the fourth figure is under five ignore it, and if it is five or over add one to the preceding figure. (If you have problems with this, refer now to Chapter 2.)

To change a fraction into its decimal equivalent divide the numerator by the denominator of the fraction to be converted.

Examples

1 Convert $\frac{1}{4}$ into a decimal: $1 \div 4 = 0.25$
2 Convert $\frac{2}{5}$ into a decimal: $2 \div 5 = 0.4$
3 Convert $\frac{2}{9}$ into a decimal: $2 \div 9 = 0.22222222... = 0.222$
4 Convert $2\frac{3}{8}$ into a decimal: $3 \div 8 = 0.375; \quad 2 + 0.375 = 2.375$
5 Convert $5\frac{8}{9}$ into a decimal: $8 \div 9 = 0.88888...; \quad 4 + 0.889 = 4.889$

Exercise 4.5

Convert the following fractions into their decimal equivalents

(a) $1\frac{3}{4}$
(b) $3\frac{5}{8}$
(c) $2\frac{7}{8}$
(d) $4\frac{1}{5}$

| 4.8 | Decimals to fractions |

In Chapter 2 we looked fairly closely at the meaning of figures in a number written in decimal notation. Briefly we said that the figures before the decimal point were units, tens, hundreds or thousands, etc., and the first figure after the decimal point, tenths, the second hundredths and the third thousandths (if you have a problem with this look back to Chapter 2).

Example

If we look at a number, say 0.375, then it can be said that it consists of 3 tenths,

7 hundredths and 5 thousandths, and can be written as

$$0.375 = \frac{3}{10} \div \frac{7}{100} + \frac{5}{1000}$$

Find the lowest common denominator.

$$= \frac{300}{1000} + \frac{70}{1000} + \frac{5}{1000}$$

1,000 is the LCD.

$$= \frac{375}{1000}$$

Although the answer $\frac{375}{1000}$ is correct we should look to see if the answer can be reduced to a more manageable fraction. This can be achieved by dividing the numerator and the denominator by the highest common factor, as we did on page 32. If, however, such a number cannot be easily found, as in this case, the fraction can be reduced using a smaller number that will divide readily into the top and the bottom of the fraction.

The number 5 will divide into 375 and 1,000, so I shall use it as my starting point.

$$\frac{375}{1000} \div \frac{5}{5} = \frac{75}{200}$$

The fraction can be reduced by dividing by 5.

$$\frac{75}{200} \div \frac{5}{5} = \frac{15}{40}$$

And again.

$$\frac{15}{40} \div \frac{5}{5} = \frac{3}{8}$$

And again.

We are now left with a much easier fraction, $\frac{3}{8}$, than the one we started with, $\frac{375}{1000}$. However, it should be noted that not all large fractions can be reduced and some, although they may not look it, will already be in their lowest form.

The above example demonstrated how to change a decimal into a fraction. Notice that we ended up with 375 over 1,000 as a fraction, and all we did was to cancel it down to its lowest fraction, $\frac{3}{8}$. This gives us three rules for changing decimal numbers into fractions:

1. *The numerator of the fraction is the decimal number to be converted, with the decimal point removed.*
2. *The denominator of the fraction is a 1 followed by as many zeros as there are decimal places in the decimal number to be converted.*
3. *Simplify the fraction if possible.*

Examples

1 $0.4 = \frac{4}{10} \div \frac{2}{2} = \frac{2}{5}$
2 $2.24 = 2\frac{24}{100} = 2 + (\frac{24}{100} \div \frac{4}{4} = \frac{6}{25}) = 2\frac{6}{25}$

Some of the more common fractions and their decimal equivalents are:

$$\frac{1}{10} = 0.1, \frac{1}{5} = 0.2, \frac{1}{3} = 0.333, \frac{2}{5} = 0.4, \frac{1}{2} = 0.5, \frac{1}{4} = 0.25, \frac{3}{4} = 0.75, \frac{1}{8} = 0.125,$$
$$\frac{3}{8} = 0.375, \frac{5}{8} = 0.625, \frac{7}{8} = 0.875, \frac{1}{100} = 0.01, \frac{1}{25} = 0.04, \frac{1}{20} = 0.05.$$

You should add to the list as you use different fractions and decimals.

Answers to exercises

4.1   (a) $5\frac{3}{8}$; (b) $3\frac{1}{18}$; (c) $10\frac{2}{15}$; (d) $8\frac{7}{8}$

4.2   (a) $1\frac{3}{16}$; (b) $\frac{7}{9}$; (c) $\frac{5}{8}$; (d) $\frac{23}{60}$

4.3   (a) $5\frac{1}{4}$; (b) $2\frac{13}{16}$; (c) $4\frac{5}{32}$; (d) $3\frac{9}{32}$

4.4   (a) $2\frac{1}{3}$; (b) $1\frac{4}{5}$; (c) $3\frac{5}{17}$; (d) $1\frac{1}{5}$

4.5   (a) 1.75; (b) 3.625; (c) 2.875; (d) 4.2

Summary

We have been looking at:

- Proper fractions.
- Improper fractions.
- Multiplying, dividing, adding and subtracting fractions.
- Converting fractions to decimals and decimals to fractions.

Things to remember

The number on the bottom of the fraction is called the denominator; if this is larger than the figure on top, called the numerator, the fraction is proper.

If the numerator is larger than the denominator the fraction is said to be improper.

Improper fractions can be turned into mixed numbers.

Adding fractions $\qquad \dfrac{a}{b} + \dfrac{c}{d} = \dfrac{a \times d + c \times b}{b \times d}$

Subtracting fractions $\qquad \dfrac{a}{b} - \dfrac{c}{d} = \dfrac{a \times d - c \times b}{b \times d}$

Multiplying fractions $\qquad \dfrac{a}{b} \times \dfrac{c}{d} = \dfrac{a \times c}{b \times d}$

Dividing fractions $\qquad \dfrac{a}{b} \div \dfrac{c}{d} = \dfrac{a}{b} \times \dfrac{d}{c} = \dfrac{a \times d}{b \times c}$

Fractions to decimals

Divide the numerator by the denominator of the fraction to be converted.

Converting decimals to fractions

The numerator of the fraction is the decimal number to be converted multiplied by 10, 100 or 1,000, etc., to produce a whole number. The denominator of the fraction will be 10, 100 or 1,000, etc., depending on which was used to produce the numerator. Finally, simplify the fraction if possible.

Take a break from your studies before testing your skills on the following self-assessment questions.

Self-assessment questions

Below, there are sixteen questions. Take your time in answering them.

- These are not meant as a test, the questions are simply to help you learn.
- Look back at your own notes and Chapter 4 if you need help.
- Answers and comments follow the questions, but you should look at them only when you have finished or are really stuck.

SAQ 4.1   Is it true that $\frac{3}{8} = \frac{6}{16}$?

SAQ 4.2   Express $\frac{16}{40}$ in its lowest term.

SAQ 4.3   Change $\frac{30}{12}$ into a mixed number.

SAQ 4.4   What is $3\frac{1}{4} + 2\frac{7}{8}$?

SAQ 4.5   What is $2\frac{1}{2} + 1\frac{1}{4} + 2\frac{3}{8}$?

SAQ 4.6   What is $2\frac{2}{3} - 1\frac{1}{4}$?

SAQ 4.7   What is $2\frac{1}{4} - \frac{9}{16}$?

SAQ 4.8   What is $1\frac{3}{4} - 1\frac{2}{3} + 1\frac{1}{4}$?

SAQ 4.9   What is $\frac{4}{5} \times 2\frac{1}{2}$?

SAQ 4.10   What is $1\frac{1}{2} \times 1\frac{1}{3} \times 1\frac{1}{4}$?

SAQ 4.11   What is $3\frac{1}{4} \div \frac{1}{4}$?

SAQ 4.12   What is $1\frac{4}{5} \div \frac{7}{10}$?

SAQ 4.13   Convert $1\frac{3}{4}$ to a decimal.

SAQ 4.14   Convert $27\frac{9}{16}$ to a decimal.

SAQ 4.15   Convert 3.7 to a fraction.

SAQ 4.16   Convert 1.45 to a fraction.

Answers and comments

SAQ 4.1

Yes, it is. You should remember, however, that the fraction is easier to work with, both mathematically and when used for measurement, when expressed in its lowest term. This is the reason we change it from $\frac{6}{16}$ to $\frac{3}{8}$. Both represent the same amount.

SAQ 4.2

The highest number that the numerator and denominator can be divided by is 8. Therefore $\frac{16}{40}$ becomes $\frac{2}{5}$.

SAQ 4.3

30 divided by 12 goes twice, with six remaining, giving $2\frac{6}{12}$. Reducing the fraction to its lowest term gives the answer $2\frac{1}{2}$.

SAQ 4.4

Adding the whole numbers together, finding the lowest common denominator, and multiplying each numerator by the number of times its denominator goes into the LCD gives us $5\frac{2+7}{8}$. Adding the numerators gives us $5\frac{9}{8}$ and reducing this to its lowest term gives the answer $6\frac{1}{8}$.

SAQ 4.5

There are two methods that can be used to solve this problem. The first is to add the first two numbers together and add the third to the result. This is useful if there is not an obvious LCD. The second is the method we used in the example in the text, and is the method we shall use now.

1. Add the whole numbers together and find the LCD $= 5\frac{4+2+3}{8}$.
2. Complete the sum and reduce to lowest term $= 5\frac{9}{8} = 6\frac{1}{8}$.

(*Note*: It is worth noting that a common denominator can always be found by simply multiplying all the denominators together. While this method will not necessarily produce the *lowest* common denominator, it will at least get you started.)

SAQ 4.6

Take away the whole numbers and find the LCD, $(1\frac{8-3}{12})$. Complete the sum and reduce to lowest term $(1\frac{5}{12})$.

SAQ 4.7

Find the LCD, $(2\frac{4-9}{16})$. To allow us to take 9 from 4 we have to rearrange the figures. Bearing in mind that the whole number 2 is made up of two lots of sixteenths, we can borrow one lot to help us. The calculation will now look like $1\frac{20-9}{16}$, which, when brought to its lowest term, will give the answer $1\frac{11}{16}$.

SAQ 4.8

Take away and add up the whole numbers and find the LCD, $(1\frac{9-8+3}{12})$. Complete the calculation and reduce to its lowest term $(1\frac{4}{12} = 1\frac{1}{3})$. An alternative method would be $1\frac{3}{4} + 1\frac{1}{4} = 3$ and $3 - 1\frac{2}{3} = 1\frac{1}{3}$.

SAQ 4.9

Change mixed numbers to improper fractions and cross cancel $(\frac{4}{5} \times \frac{5}{2} = \frac{2}{1} \times \frac{1}{1})$. Multiply numerators and denominators, and reduce to its lowest term $(\frac{2}{1} = 2)$.

SAQ 4.10

Change mixed numbers to improper fraction and cross cancel $(\frac{3}{2} \times \frac{4}{3} \times \frac{5}{4} = \frac{1}{2} \times \frac{1}{1} \times \frac{5}{1})$. Multiply numerators and denominators, and reduce to its lowest term $(\frac{5}{2} = 2\frac{1}{2})$.

SAQ 4.11

All mixed numbers are turned into improper fractions $(\frac{13}{4} \div \frac{1}{4})$. The divisor is inverted and the division sign is changed to a multiplication sign $(\frac{13}{4} \times \frac{4}{1})$. Cross cancel where possible $(\frac{13}{1} \times \frac{1}{1}$, the 4s cancel out). Multiply numerators and denominators and reduce to lowest term $(\frac{13}{1} = 13)$.

SAQ 4.12

All mixed numbers are turned into improper fractions $(\frac{9}{5} \div \frac{7}{10})$. The divisor is inverted and the division sign is changed to a multiplication sign $(\frac{9}{5} \times \frac{10}{7})$. Cross cancel where possible $(\frac{9}{1} \times \frac{2}{7}$, the 5 and the 10 cancel out). Multiply numerators and denominators and reduce to lowest term $(\frac{18}{7} = 2\frac{4}{7})$.

SAQ 4.13

$1\frac{3}{4}$ written as a decimal $= 1 + (3 \div 4) = 1 + 0.75 = 1.75$.

SAQ 4.14

$27\frac{9}{16}$ expressed as a decimal $= 27 + (9 \div 16) = 27 + 0.5625$. Working to three decimal places $= 27.563$.

SAQ 4.15

3.7 expressed as a fraction $= 3\frac{7}{10}$. The 0.7 represents the value of the tenths in the number.

SAQ 4.16

1.45 written as a fraction $= 1\frac{45}{100}$. This written in its lowest term is $1\frac{9}{20}$, where the numerator and denominator have both been divided by 5.

5.1 Measurement

Measurement of length or linear measurement is used on a daily basis by most people employed in the building industry. Technically speaking, a length measurement has only one dimension; that is, it has no width or depth. We use the term to describe the distance from one point to another. Whatever is used to obtain a measurement, whether it be a length of string with a knot in it or a 10 m tape, both are measuring in units (the string, from one end to the knot, would be one unit).

Providing you do not need to pass the information on to someone who does not understand how your 'system' works, string, wire or a length of wood can be very accurate methods of transferring measurement, especially in repetitive work where consistent accuracy is called for.

Gauge rods or storey rods are good examples; bricklayers use these when building brick walls (Figure 5.1). The rods are usually made at the site from 40 mm timber. On the rod the bricklayer marks the height of the brick courses and uses the rod to check that the work is keeping to the correct gauge. The rod can be extended to the height of one storey and marked with cill heights, lintel heights and floor heights.

Gauges of this sort are used successfully every day in a variety of jobs both connected and unconnected with the building trade. They work fine providing, as I said earlier, everyone who deals with the gauge is aware of the units; if not, a standard unit must be used.

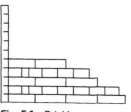

Fig. 5.1 Bricklayers gauging rod

5.2 Standard units

There are two standard measurement systems in use today: the metric system and the British Imperial system. Both are used to express the amount or the size of something and its physical properties.

The majority of measurements taken from drawings or architects' instructions, textbooks or information from college will be given using the metric system. However, the Imperial system is still in use and is often mixed with the metric system, particularly on building sites or at builders' merchants. To help with this problem, this chapter includes some common conversions from the metric system to the Imperial system and from the Imperial system to the metric system.

5.3 Metric units

The main advantage of the metric system lies in its simplicity. It is, like our number system, a decimal system – that is, it has a base of ten. The multiple units or larger units are multiples of ten of the base unit and the sub-units or smaller units are decimal fractions of the base units.

The metric system has been redefined since being introduced in Europe in the nineteenth century and is perhaps now better known as the Système International d'Unités, or simply SI. The SI differs from the traditional metric system in that it recommends only some of the metric units and their multiples and sub-multiples for use.

When making measurements of a physical quantity, such as a piece of wood, the final answer must be expressed as a number, followed by the unit or units that are being used. The units depict the standard that is being used, i.e. metres, kilograms, seconds, etc., while the number relates to the amount of the units, i.e. 3 metres, 5 kilograms, 30 seconds, etc.

There are, for our purposes, two kinds of units: **base** and **derived**. Below are listed the units with which readers should familiarise themselves.

SI base units

Quantity	Unit	Unit symbol
length	metre	m
mass (see note)	kilogram	kg
time	second	s
temperature	kelvin	K
electrical current	ampere	A
luminous intensity	candela	cd

If you measure a length of wood your answer will be in metres or parts of a metre; the unit symbol 'm' would be placed after the number to tell everyone the standard being used.

Note: The word **mass** is used in the SI in preference to **weight**. Weight describes a pulling force (gravity) the Earth exerts on objects. The strength of this force depends on the object's position in relation to the centre of the Earth. The nearer it is to the centre of the Earth the greater will be the gravitational force; an object a great distance from the centre of the Earth, e.g. in outer space, will become increasingly less influenced by gravity, and at very great distances it is said to be *weightless*.

Mass is a fixed property of an object and depends on the internal structure of the object and its size. As these will not change, irrespective of the object's position, it is a more accurate measurement. Although the term mass is used throughout this book, the term weight is not expected to disappear from general usage.

Derived units

Derived units can be formed by combining any number of base units, for example the unit of force is the newton, which is produced by combining the first three base units of length, mass and time to obtain kg m/s^2.

Some of the more common derived units used in the building trade are shown below:

Quantity	Unit	Unit symbol
area	square metre	m^2
volume (see note)	cubic metre	m^3
capacity (see note)	litre	litre
force	newton	N
density = mass/volume	kilogram per cubic metre	kg/m^3
pressure	newtons per square metre	N/m^2
energy or work	joule	J
power	watt	W
frequency	hertz	Hz
electrical potential	volt	V
thermal conductivity	watts per kelvin	W/K

Note: SI makes no distinction between **capacity** and **volume**, as both relate to the same three-dimensional space. The volume is the space an object occupies, while capacity is the amount that the space can contain.

5.4   Multiples and sub-multiples of SI units

No single unit is a convenient size for all the measurements you are likely to make. The thickness of a mortar joint is very small when compared with the whole wall and even smaller when set against the building site. To use the same unit of measurement (the metre) for all three would be complicated, requiring lots of noughts or powers. Instead we use prefixes to produce new units.

There are at least sixteen prefixes that can be attached to a unit to produce multiples and sub-multiples of units. Below are the prefixes and the symbols most used in the building trade.

Multiplication factor	Prefix	Symbol
$1,000,000 = 10^6$	mega	M
$1,000 = 10^3$	kilo	k
$100 = 10^2$	hecto	h
$0.01 = 10^{-2}$	centi	c
$0.001 = 10^{-3}$	milli	m
$0.0001 = 10^{-6}$	micro	μ

- All prefixes must be written in small type, as must the symbols abbreviating them, the exception being 'M' (mega).
- The word mega or the symbol 'M' in front of a unit requires that the unit be increased by a million, e.g. 1 megajoule = 1 million joules.
- The prefix kilo or the symbol 'k' in front of a unit represents an increase of a thousand, e.g. 1 kilogram = 1,000 grams.
- The prefix hecto* or the symbol 'h' in front of a unit represents an increase of one hundred, e.g. 1 hectolitre = 100 litres.
- The prefix centi* or the symbol 'c' in front of a unit represents a decrease of a hundred, e.g. 1 centimetre = 0.01 of a metre or $\frac{1}{100}$ of a metre.
- The prefix milli or the symbol 'm' in front of a unit requires that the unit be decreased by a thousand, e.g. a millimetre = 0.001 of a metre or $\frac{1}{1000}$ of a metre.
- The prefix micro* or the symbol (μ the twelfth letter of the Greek alphabet) in front of a unit represents a decrease of 1 million, e.g. 1 microgram = 0.000001 of a gram or $\frac{1}{1000000}$ of a gram.

Note: The prefixes hecto, centi and micro are rarely used on the practical side of the building trade but are regularly found in textbooks and test papers.

We can now talk about the mortar joint being so many millimetres (mm) thick. The wall being so many metres (m) long and the building site can be expressed in either metres (m) or, if it is large enough, in kilometres (km).

5.5 Metric/Imperial conversions

In any system of measurement three fundamental units are required. These are units of **length**, **mass** and **time**. The base units for the SI system are the **metre** (m), **kilogram** (kg) and **second** (s). In the Imperial measurement system the standard units of length and mass are the **foot** (ft) and the **pound** (lb). The unit of time is the second, which is the same in the SI. The gallon is also defined in the Imperial system as the volume occupied by exactly 10 pounds of water at a set density.

Metric to Imperial	**Imperial to Metric**
Length	*Length*
1 millimetre = 0.039 inch	1 inch = 25.4 millimetres
1 centimetre = 0.3937 inch	1 foot = 0.3048 metre
1 metre = 39.37 inches or 1.09 yards	1 yard = 0.9144 metre
1 kilometre = 0.6214 mile	1 mile = 1.609 kilometres

Area
1 square millimetre =
0.00155 square inch

1 square centimetre =
0.155 square inch
1 square metre =
10.76 square feet
or 1.196 square yards
1 hectare (see note 1) = 2.471 acres

Volume
1 cubic millimetre =
0.0061 cubic inch
1 cubic centimetre =
0.061 cubic inch
1 cubic metre = 1.308 cubic yards

Capacity
1 litre = 1.76 pints
1 litre = 0.21997 gallon
1 litre = 61.024 cubic inches

Mass
1 kilogram = 2.205 pounds

Area
1 square inch =
645.2 square millimetres
or 6.452 square centimetres
1 square foot = 0.093 square metre

1 square yard = 0.836 square metre

1 acre = 0.405 hectare

Volume
1 cubic inch =
16,393 cubic millimetres
1 cubic foot = 0.028316 cubic metre

1 cubic yard = 0.765 cubic metre

Capacity
1 pint = 0.568 litre
1 gallon = 4.546 litres

Mass
1 pound = 0.454 kilogram
1 ton (see note 2) (2,240 lb) =
1,016 kilograms

Note 1: The SI unit of area is the square metre (m^2), but for large areas of land measurement the hectare is more normally used. A hectare is equal to 1,000 square metres.
Note 2: 1 tonne = 1,000 kilograms.

Temperature

- To convert Fahrenheit to Celsius (Centigrade), subtract 32 and divide by 1.8.
- To convert Celsius to Fahrenheit, multiply by 1.8 and add 32.

Using the conversion table

Do not feel it necessary to commit all the exact conversions to memory. You should, however, look at the conversions and find the ones that are peculiar to or are most often used in your trade (if you are not sure ask your lecturer for help).

Many of the conversions can be rounded off or shortened for everyday use on a building site. For example, a painter would find it only necessary to remember that there are approximately 4.5 litres in a gallon of paint rather than exactly 4.546, and bricklayers that a cubic yard of sand is approximately 0.75 cubic metre of sand.

To use the conversions a multiplication calculation is necessary. Below are some examples that we can work through together.

Examples

1 Convert 11 feet into metres.

From the list of conversions it can be seen that 1 foot is equal to 0.3048 metre, and as we have 11 feet, this is

$$0.3048 \times 11 = 3.3528 \text{ metres}$$

which for our purposes we can round off to 3.353 m

2 How many cubic metres of sand are there in 3 cubic yards? There is 0.765 cubic metre in a cubic yard, therefore the number of cubic metres is:

$$0.765 \times 3 = 2.295 \text{ m}^3$$

3 How many miles would 120 kilometres represent?

There is 0.6214 mile in a kilometre, therefore the number of miles would be:

$$120 \times 0.6214 = 74.568 \text{ miles}$$

4 If the temperature was given as 77° Fahrenheit, what would it be using the Celsius scale?

Using the formula on the conversion page, deduct 32 from 77 and divide the answer by 1.8:

$$77 - 32 = 45; \ 45 \div 1.8 = 25 \ °C$$

Summary

We have been looking at:

- The use of gauges.
- Measurement and standard units.
- Metric and Imperial units.
- Converting units.

Things to remember

- Gauges are an accurate method of transferring sizes. It is important, however, to ensure that everyone who comes into contact with your gauge is aware of the units being used.
- Standard units should be used for passing on measurements, preferably units from the metric system.
- The metric system (SI) is made up of base and derived units and their multiples and sub-multiples.

- Conversion from one system to another is possible using approximations; however, some accuracy is lost.

Take a break from your studies now. There are no self-assessment questions for you to answer in this chapter. Instead, practise the conversions and measuring elements using the SI.

Squares, powers and scientific notation

> *Prior knowledge* If you feel you are already skilled in the subjects mentioned above, turn to page 62 and try working through the SAQs. If you find difficulty in answering them, work through this chapter.

6.1 Squares

Look again at the multiplication table (Figure 6.1, overpage) first used in Chapter 1.

Notice this time that the numbers 4, 9, 16, 25, 36, 49, 64, 81, 100, 121, 144, 169, 196, 225, 256, 289 and 324 are in a heavier type than the rest of the numbers in the table. These numbers are the result of multiplying 2 with 2, 3 with 3, 4 with 4 and so on up to 18 with 18. The answer you obtain when you multiply a number by itself is called the **square** of that number. For example, 64 is the square of 8 ($8 \times 8 = 64$). This is also called raising a number to its **second power** and is written using a small 2 above the number like this: 8^2.

Exercise 6.1

(a) Using Figure 6.1, find the square of 14.
(b) What is 12 raised to its second power?
(c) What is 7^2?

On your calculator you should have a button marked $[x^2]$. If you enter a number and press $[x^2]$ the display will change to the square of that number. Try it now. If you have a calculator that does not provide an $[x^2]$ button, the square of a number can still be found (remember: the square of a number is that number multiplied by itself). If you are asked to find the square of 21,

1	2	3	4	5	6	7	8	9	10	11	12	13	14	15	16	17	18
2	4	6	8	10	12	14	16	18	20	22	24	26	28	30	32	34	36
3	6	9	12	15	18	21	24	27	30	33	36	39	42	45	48	51	54
4	8	12	16	20	24	28	32	36	40	44	48	52	56	60	64	68	72
5	10	15	20	25	30	35	40	45	50	55	60	65	70	75	80	85	90
6	12	18	24	30	36	42	48	54	60	66	72	78	84	90	96	102	108
7	14	21	28	35	42	49	56	63	70	77	84	91	98	105	112	119	126
8	16	24	32	40	48	56	64	72	80	88	96	104	112	120	128	136	144
9	18	27	36	45	54	63	72	81	90	99	108	117	126	135	144	153	162
10	20	30	40	50	60	70	80	90	100	110	120	130	140	150	160	170	180
11	22	33	44	55	66	77	88	99	110	121	132	143	154	165	176	187	198
12	24	36	48	60	72	84	96	108	120	132	144	156	168	180	192	204	216
13	26	39	52	65	78	91	104	117	130	143	156	169	182	195	208	221	234
14	28	42	56	70	84	98	112	126	140	154	168	182	196	210	224	238	252
15	30	45	60	75	90	105	120	135	150	165	180	195	210	225	240	255	270
16	32	48	64	80	96	112	128	144	160	176	192	208	224	240	256	272	288
17	34	51	68	85	102	119	136	153	170	187	204	221	238	255	272	289	306
18	36	54	72	90	108	126	144	162	180	198	216	234	252	270	288	306	324

Fig. 6.1

try putting in the number 21 press [×] and [=] and the answer should be 441, i.e. $21^2 = 441$.

 Exercise 6.2

(a) Use your calculator to check that the rest of the table in Figure 6.1 is correct.
(b) What is 17 raised to its second power?
(c) What is 4.5^2.

6.2 Square roots

You should also have on your calculator a button marked [$\sqrt{}$]. This is used for finding the **square root** of numbers. The square root of a number is that value which, when multiplied by itself, gives the original number. It might be easier to think of it as the opposite of squaring a number. For example, the square of 8 is 64 and the square root of 64 is 8.

If you were writing down the square root of 64, the 64 would be placed within the square root sign $\sqrt{64}$. This shows that it is the square root of 64 that is required.

Exercise 6.3

(a) Use your calculator to find the square root of 100.
(b) What is the square root of 256?
(c) Find $\sqrt{441}$.

The method for working out square roots without a calculator facility is complicated and is beyond the scope of this book.

6.3 Powers

Above, we looked at squaring numbers or raising them to the power of 2, both of which mean the same thing. We learnt that the small 2 above a number required that number to be multiplied by itself or squared.

Look at the calculation $6 \times 6 \times 6 \times 6 \times 6$. Is there an easier method of expressing it? (Remember that 6×6 can also be written as 6^2.) It follows then that $6 \times 6 \times 6 \times 6 \times 6$ can be written as 6^5.

The 6 is called the **base** and the 5 the **exponent** or **index** (plural *indices*) of the power and it would be read as the fifth power of six or, more commonly, six to the power of five.

Powers enable us to write repeated operations in this quicker way, and makes talking about and manipulating large numbers easier.

Some calculators have a button marked [x^y], where x represents the base number and y the exponent. If you have a calculator with this facility try it, press [6] then [x^y] then [5] then [=]. You can check your answer by working out the sum $6 \times 6 \times 6 \times 6 \times 6$. Both methods should give the same answer 7,776; see which you think is quickest.

Exercise 6.4

Give the value of:

(a) 6^5
(b) 16^3
(c) 3^{12}
(d) 10^5
(e) 3.1^3
(f) 2.7^2

Note: It is worth noting that x^2 is normally referred to as 'x squared' and x^3 as 'x cubed', whereas powers greater than three are known as 'x to the power of, – for example, x^4 is known as 'x to the power of four'.

6.4 Multiplying and dividing by 10

Look again at the columns we first used in Chapter 3 to learn about place values for numbers:

100,000	10,000	1,000	100	10	1	•	0.1	0.01	0.001

We said that each column represented ten times the value of the column to its right. Therefore, multiplying a number by 10 simply involves moving each figure one column to the left.

Example

$935.62 \times 10 =$

100,000	10,000	1,000	100	10	1	•	0.1	0.01	0.001
			9	3	5	•	6	2	

Move one column to the left =

100,000	10,000	1,000	100	10	1	•	0.1	0.01	0.001
		9	3	5	6	•	2		

9,356.2

Multiplying by 100 is really multiplying by 10, and then by 10 again, so the figures move two places to the left.

Example

$439.2 \times 100 =$

100,000	10,000	1,000	100	10	1	•	0.1	0.01	0.001
			4	3	9	•	2		

Move two columns to the left =

100,000	10,000	1,000	100	10	1	•	0.1	0.01	0.001
	4	3	9	2	0	•			

43920

A 0 is placed after the 2 to tell us there are no units and to keep the other numbers in their correct place. And since dividing by 10 is the opposite of multiplying by 10, we have the corresponding rule: to divide by 10, move all the numbers one place to the right.

Example

$354.57 \div 10 =$

100,000	10,000	1,000	100	10	1	•	0.1	0.01	0.001
			3	5	4	•	5	7	

Move one column to the right =

100,000	10,000	1,000	100	10	1	•	0.1	0.01	0.001
				3	5	•	4	5	7

35.457

Some examples of multiplying and dividing by 10

$31.07 \times 10 = 310.7$ $0.003 \div 10 = 0.0003$
$54 \times 10 = 540$ $3875 \div 10 = 387.5$
$0.008 \times 10 = 0.08$ $846 \div 10 = 84.6$
$8.001 \times 10 = 80.01$ $6.943 \div 10 = 0.6943$

Look carefully at the examples, making sure you understand what has been done, and then try Exercise 6.5.

Exercise 6.5

Give the value of:

(a) 24.576×10
(b) 0.121×10
(c) $7,444.23 \times 10$
(d) $43.0003 \div 10$
(e) $0.45322 \div 10$
(f) $0.00023 \div 10$

6.5   Multiplying and dividing by powers of 10

Often when working on calculations, you will be required to multiply or divide by 10 a certain number of times. For example, if you are asked to multiply 34.567 by 1,000 one way of thinking about and writing down the figures would be as $34.567 \times 10 \times 10 \times 10$. However, this can be quite laborious, especially if you have been asked to multiply by a much larger number, say two million. But as ever there is an easier way.

Think back to the previous page, and the subject of powers and what we learnt about repeated operations. It was said that 6^5 was a much easier method of writing $6 \times 6 \times 6 \times 6 \times 6$ and that both meant the same thing. So it follows, then, that a better way of writing the calculation

$34.567 \times 10 \times 10 \times 10$ would be 34.567×10^3. (This method of writing numbers is called **scientific notation**.) The index 3 indicates that we have applied the operation 'multiply by 10' three times (i.e. 10^3 or ten cubed). The overall effect of this would be to move all the figures three places to the left, giving the answer 34,567.

We can treat division similarly. For example, if we are asked to divide 34.567 by 1,000 the calculation can be thought of as $34.567 \div 10 \div 10 \div 10$; however, this time we are *dividing* by 10 rather than *multiplying* by 10, so we move the figures three places to the *right*, giving the result

$$34.567 \div 1,000 = 0.034567$$

or better still

$$34.567 \div 10^3 = 0.034567$$

Now, since when we divide by 10 we move the figures in the reverse direction to when we multiply by 10, there is another way of writing down what we have done:

$$34.567 \times 10^{-3} = 0.034567$$

The '10^{-3}' should be interpreted as an instruction to move the figures three places to the right. (*Note*: It is normal practice that 10^{-3} is referred to as 'ten to the power of minus three', rather than 'ten minus cubed'.) So for our purposes we can now say:

- '$\times\ 10^3$' means apply the operation 'multiply by 10 three times' or move the figures three places to the left.
- '$\times\ 10^{-3}$' means apply the operation 'divide by 10 three times' or move the figures three places to the right.

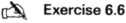 **Exercise 6.6**

Complete the following:

(a) $73.239 \times 10^{-3} =$
(b) $27.4 \times 10^4 =$
(c) $0.79 \times 10^{-3} =$
(d) $7.7 \times 10^0 =$
(e) $0.003 \times 10^1 =$
(f) $93.724 \times \qquad = 93,724$
(g) $7.321 \times \qquad = 0.7321$
(h) $327 \times \qquad = 327,000$
(i) $0.24 \times \qquad = 0.000024$
(j) $0.0009 \times \qquad = 90$

You will have probably noticed that multiplying by 10^3 is the same as multiplying by 1,000; that multiplying by 10^1 is the same as multiplying by

10 itself; that multiplying by 10^0 is the same as multiplying by 1; and, possibly, that multiplying by 10^{-1} is the same as multiplying by 0.1.

These 'powers of 10' can be used as an alternative way of writing these numbers, e.g.

$$10^3 \quad = \quad 1,000$$
$$10^1 \quad = \quad 10$$
$$10^0 \quad = \quad 1$$
$$10^{-1} = \quad 0.1$$

Exercise 6.7

Use the pattern that suggests itself to complete the table below:

$$10^6 \quad = \quad 1,000,000$$
$$10^5 \quad = \quad\quad 100,000$$
$$10^4 \quad =$$
$$10^3 \quad = \quad\quad\quad 1,000$$
$$10^2 \quad =$$
$$10^1 \quad = \quad\quad\quad\quad 10$$
$$10^0 \quad = \quad\quad\quad\quad 1$$
$$10^{-1} = \quad\quad\quad\quad 0.1$$
$$10^{-2} =$$
$$10^{-3} =$$
$$10^{-4} =$$
$$10^{-5} =$$
$$10^{-6} =$$

6.6 Multiplying and dividing powers

Powers of the same base can be multiplied and divided providing some simple rules are adhered to. Let us look at two examples:

(a) $3^2 \times 3^5 =$
(b) $6^5 \div 6^2 =$

Use your calculator to evaluate (a). Your answer should have been 2,187. Think about the power to which 3 must be raised to obtain 2,187 and how, using your calculator, you can find this out.

Enter 3 on your calculator and multiply by 3. Repeat the operation until the number 2,187 is displayed, counting the number of times you multiply as you go. You should find that you multiplied by 3 six times before your calculator displayed the value, i.e.

$$3 \times 3 \times 3 \times 3 \times 3 \times 3 = 2,187$$

As we began with 3 and then multiplied by 3 six times, we can say we raised 3 to the seventh power. So

$$3^2 \times 3^5 = 3^7$$

(writing the expression $3^2 \times 3^5$ as 3^7 is called **simplifying**)

Looking at the result for (a), you can see what has happened:

$$3^2 \times 3^5 = 3^{2+5} = 3^7$$

The indices have been added. So we can now say that *providing the bases are the same, numbers raised to a given power can be* **multiplied** *together by* **adding** *their indices, i.e.* $a^x \times a^y = a^{x+y}$.

Now try (b). I shall give you the explanation, but you should try it yourself in the same way as above.

$$6^5 \div 6^2 = 6^{5-2} = 6^3$$

The indices have been subtracted. So we can now say that *providing the bases are the same, numbers raised to a given power can be* **divided** *by each other by* **subtracting** *their indices, i.e.* $a^x \div a^y = a^{x-y}$.

Exercise 6.8

Simplify the following:

(a) $4^3 \times 4^4 =$
(b) $10^5 \times 10^3 =$
(c) $5^8 \div 5^3 =$
(d) $10^7 \div 10^2 =$

6.7 Scientific notation

Scientific notation is an important method of writing, in a simpler form, both very large and very small numbers. Many calculators use scientific notion to save space on the display area. It uses a method of writing numbers as the product of a number lying between 1 and 10, and a power of 10.

If your calculator does this you can see what I mean by entering in a number, say 45.678, and repeatedly multiplying by 10. On a calculator with an 8-digit display, what happens after you reach 45,678,000 and multiply again by 10? Most calculators will display the result as $4.5678\ ^{08}$ or something similar. This is intended to mean 4.5678×10^8 or 456,780,000.

Repeat the above procedure, but this time investigate what happens when you repeatedly divide 45.678 by 10. The display should be something like $4.5678\ ^{-08}$. Again this is shorthand for 4.5678×10^{-8} or 0.00000004568.

6.8 Multiplying and dividing numbers in scientific notation

To multiply numbers written in scientific notation, multiply the numbers together and add the indices, i.e.

$$(4 \times 10^3) \times (3 \times 10^5) = (4 \times 3) \times (10^3 \times 10^5)$$
$$= (4 \times 3 = 12) \times (10^3 \times 10^5 = 10^8)$$
$$= 12 \times 10^8 \quad \text{(Make sure you have followed this exactly)}.$$

To express this in scientific notation, 12×10^8 must be expressed as a product of a number between 1 and 10. So the 12 is divided by 10 and the 10^8 multiplied by 10, to give 1.2×10^9.

Note: Remember that dividing or multiplying by 10 is the same as moving the numbers one place to the left or right. For example, had the answer been 0.012×10^8 then the number point would move two places to the left and two would be deducted from the indices giving 1.2×10^6.

To divide numbers in scientific notation, divide the numbers into each other and subtract the indices, i.e.

$$(6 \times 10^8) \div (3 \times 10^3) = (6 \div 3) \times (10^8 \div 10^3)$$
$$= (6 \div 3 = 2) \times (10^8 \div 10^3 = 10^5)$$
$$= 2 \times 10^5$$

Once again ensure you have understood what I have done and then try the exercise.

Exercise 6.9

Calculate the following, giving your answers in scientific notation:

(a) $(5 \times 10^2) \times (5 \times 10^3)$
(b) $(16 \times 10^9) \times (4 \times 10^2)$
(c) $(8 \times 10^{12}) \div (2 \times 10^6)$
(d) $(6 \times 10^8) \div (2 \times 10^4)$

6.9 Significant figures

In Chapter 2 we learnt about rounding off numbers that were correct to so many decimal places. We said that the accuracy required in the answer usually decided the number of decimal places we worked to.

The next exercise asks you to do something that may at first seem silly.

Exercise 6.10

Work out $2 \div 789$, giving your answer to two decimal places. Do not use scientific notation in your answer.

The above exercise shows that giving your answer to, say, two decimal places is not always the most sensible idea. The answer to the exercise shows that $2 \div 789 = 0.00$. If, however, we write our answer to the exercise in scientific notation, we get

$$2 \div 789 = 2.5348542 \times 10^{-3}$$

Now we can round off the 2.5348542 part of this to two decimal places, giving 2.53×10^{-3}. This way we get round the difficulty of giving an answer that does not contain too many unnecessary figures, and at the same time does not run the risk of throwing away all the information as we did in the first part of the exercise.

    The answer 2.53×10^{-3}, or written out as 0.00253, is said to be given to three significant figures. (The three significant figures are 2, 5 and 3.)

    It is not essential to write numbers in scientific notation before rounding them to three significant places. I have just done it this way to avoid some of the difficulties that sometimes arise in deciding exactly what is meant by 'significant figures'. For example, in the number 0.0000040567 the only zero that counts as a significant figure is the one between 4 and 5, the other zeros being there just to indicate the place value of the figures 4, 5, 6 and 7. Zeros only count as significant figures when they are preceded or followed by non-zero digits. Using scientific notation this number is 4.0567×10^{-6}, and when rounded off to three significant places it would become 4.06×10^{-6} (the 5 has been rounded up to 6.)

Answers to exercises

6.1   (a) 196; (b) 144; (c) 49

6.2   (a) Yes it is; (c) 289; (c) 20.25

6.3   (a) 10; (b) 16; (c) 21

6.4   (a) 7,776; (b) 4,096; (c) 531,441; (d) 100,000; (e) 29.791; (f) 7.29
    NB: The answer to (f) can be found by using either $[x^2]$ or $[x^y]$.

6.5   (a) 245.76; (b) 1.21; (c) 74,442.3; (d) 4.30003; (e) 0.045322; (f) 0.000023

6.6   (a) 0.073239; (b) 274,000; (c) 0.00079; (d) 7.7; (e) 0.03; (f) 10^3; (g) 10^{-1};
    (h) 10^3; (i) 10^{-4}; (j) 10^5

6.7 10^6 = 1,000,000
10^5 —   100,000
10^4 =   10,000
10^3 =   1,000
10^2 =   100
10^1 =   10
10^0 =   1
10^{-1} =   0.1
10^{-2} =   0.01
10^{-3} =   0.001
10^{-4} =   0.0001
10^{-5} =   0.00001
10^{-6} =   0.000001

6.8   (a) 4^7; (b) 10^8; (c) 5^5; (d) 10^5

6.9   (a) 2.5×10^6; (b) 6.4×10^{12}; (c) 4×10^6; (d) 3×10^4

Summary

We have been looking at:

- Squares and square roots.
- Powers.
- Multiplying and dividing by 10.
- Multiplying and dividing by powers of 10.
- Multiplying and dividing by powers.
- Scientific notation.
- Multiplying and dividing numbers in scientific notation.
- Significant figures.

Things to remember

- The square of a number is found by multiplying the number by itself
- Finding the square root of a number consists of finding another number which when squared equals that number. It is the opposite of squaring a number.
- Powers give an easier method of showing that a number is to be multiplied by itself. The large number, the base, is the number to be multiplied, while the small raised number, the index, indicates how many times the operation is to be carried out.
- To multiply by 10, move all the numbers one place to the left, if necessary filling in vacant spaces with zeros.
- To divide by 10, move all the numbers one place to the right, if necessary filling in vacant spaces with zeros.

- '$\times 10^x$' means apply the operation 'multiply by ten x times' or move the figures x places to the left.
- '$\times 10^{-x}$' means apply the operation 'divide by ten x times' or move the figures x places to the right.
- Providing the bases are the same, numbers raised to a given power can be multiplied together by adding their indices, i.e. $a^x \times a^y = a^{x+y}$.
- Providing the bases are the same, numbers raised to a given power can be divided by each other by subtracting their indices, i.e. $a^x \div a^y = a^{x-y}$.
- To multiply numbers written in scientific notation, multiply the first parts together and add together the powers to which the 10s are raised.
- The idea of answers being given to a set number of significant places as opposed to a set number of decimal places will often give a more accurate result.

Take a break from your studies before testing your skills on the following self-assessment questions.

Self-assessment questions

Below there are six questions. Take your time in answering them.

- These are not meant as a test; the questions are simply to help you learn.
- Look back at you own notes and Chapter 6 if you need help.
- Answers and comments follow the questions, but you should look at them only when you have finished or are really stuck.

SAQ 6.1   Define, in your own words, what is meant by the square of a number.

SAQ 6.2   What notation is used to show that (a) a number is to be squared or that (b) the square root is to be found?

SAQ 6.3   Look back to Chapter 5 if you need to and complete the following?

(a)   A kilowatt is $10^{[?]}$ watts
(b)   A kilogram is $10^{[?]}$ grams
(c)   A joule is $10^{[?]}$ megajoules
(d)   A millimetre is $10^{[?]}$ metres

SAQ 6.4   Complete the following (give your answers in terms of indices)

(a) $8^2 \times 8^3 =$
(b) $6^{2.5} \times 6^{4.5} =$
(c) $4^7 \div 4^5 =$
(d) $5^{3.5} \div 5^{1.5} =$

SAQ 6.5   Complete the following (give your answers in scientific notification)

(a) $(7 \times 10^3) \times (2 \times 10^3) =$
(b) $(3 \times 10^{2.5}) \times (4 \times 10^{1.5}) =$
(c) $(9 \times 10^5) \div (3 \times 10^2) =$
(d) $(12 \times 10^{4.5}) \div (3 \times 10^{2.5}) =$

SAQ 6.6   Write the following numbers correct (a) to two significant figures and (b) to three significant figures. The first one has been done for you.

Number	Correct to two significant figures	Correct to three significant figures
3.8072×10^5	3.8×10^5	3.81×10^5
0.003893		
3.0062×10^6		
0.00034297		
9.87124		

Answers and comments

SAQ 6.1

The square of a number is the result obtained when the number is multiplied by itself.

SAQ 6.2

(a)   A small 2 is placed above a number to be squared, i.e. $6^2 = 6 \times 6$.
(b)   A number whose square root is to be found is placed within the square root sign thus, $\sqrt{64}$.

SAQ 6.3

(a)   A kilowatt is 10^3 watts or a kilowatt is a thousand watts.
(b)   A kilogram is 10^3 grams or a kilogram is a thousand grams.
(c)   A joule is 10^{-6} megajoule or a joule is a millionth of a megajoule.
(d)   A millimetre is 10^{-3} metre or a millimetre is a thousandth of a metre.

SAQ 6.4

Using the rule 'providing the bases are the same, numbers raised to a given power can be multiplied together by adding their indices' gives us

(a) $8^2 \times 8^3 = 8^5$
(b) $6^{2.5} \times 6^{4.5} = 6^7$

Using the rule 'providing the bases are the same, numbers raised to a given power can be divided by each other by subtracting their indices' gives us

(c) $4^7 \div 4^5 = 4^2$
(d) $5^{3.5} \div 5^{1.5} = 5^2$

SAQ 6.5

To multiply numbers in scientific notation, multiply the numbers together and add the indices.

(a) $(7 \times 10^3) \times (2 \times 10^3) = 14 \times 10^6 = 1.4 \times 10^7$
(b) $(3 \times 10^{2.5}) \times (4 \times 10^{1.5}) = 12 \times 10^4 = 1.2 \times 10^5$

To divide numbers in scientific notation, divide the numbers into each other and subtract the indices.

(c) $(9 \times 10^5) \div (3 \times 10^2) = 3 \times 10^3$
(d) $(12 \times 10^{4.5}) \div (3 \times 10^{2.5}) = 4 \times 10^2$

SAQ 6.6

Number	Correct to two significant figures	Correct to three significant figures
3.8072×10^5	3.8×10^5	3.81×10^5
0.003893	0.0039	0.00389
3.0062×10^6	3.0×10^6	3.01×10^6
0.00034297	0.00034	0.000343
9.87124	9.9	9.87

Perimeters, areas and volumes

Prior knowledge If you feel you are already skilled in the subjects mentioned above, turn to page 76 and try working through the SAQs. If you find difficulty in answering them, work through this chapter.

7.1 ## Perimeters

The **perimeter** of a figure, shape or object is its boundary or outer edge. Draw a square on a piece of scrap paper. The pencil line is the boundary of the square and the square cannot go over the line unless new boundaries are drawn, thus increasing the perimeter. A football pitch is a good example of this. The white lines around the edges of the pitch denote the boundary of the playing area (the perimeter). To find the length of the perimeter measure all the sides and add them together.

The size of the sheet of A4 paper in Figure 7.1 is 210 mm × 297 mm. To find the perimeter of the paper would need the following calculation: add the four sides together

210 mm + 210 mm + 297 mm + 297 mm = 1,014 mm or 1.014 m

If the corner of the sheet were to be cut off then the perimeter would need to be recalculated, this time measuring five edges.

A carpenter, estimating the cost of skirting for a room, would first measure the length of each wall, then add the measurements together to give the amount of skirting needed, or the perimeter of the room. Multiplying the total length by the cost per metre, the carpenter would arrive at the price of the skirting required.

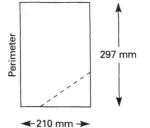

297 mm

Perimeter

◄—210 mm—►

Fig. 7.1

Example

A room measuring 6 metres by 4.5 metres will have a perimeter of

6 m + 6 m + 4.5 m + 4.5 m = 21 metres

(*Note*: In reality, an allowance would be made to this quantity for door openings and wastage, etc.) If the skirting selected is £1.50 a metre the cost will be

21 × £1.50 = £31.50

 Exercise 7.1

(a) How much skirting would be required to fit out a room measuring 7.5 metres by 5.5 metres?
(b) If the skirting cost £2.10 per metre, what would be the cost to fit out the room?
(c) If the carpenter charges 90p a metre to fit the skirting, what would be the total cost for the room?

7.2 ## Areas

Areas can be thought of as a measurement of the surface of anything. The sheet of A4 paper that we measured in Figure 7.1 has an area; it is contained or bounded by the perimeter. Builders often order materials in areas or give quotations for a specific area. Bricklayers charge so much a square metre for pointing. Prices for plastering are often quoted at so much a square metre. Most sheet materials are brought and sold in areas. They are often, in conjunction with other measurements, used to relate thickness of materials, e.g. the thickness of sheet lead is denoted by its weight. A piece of lead measuring 12 inches by 12 inches and weighing 4 lb is known as code 4 sheet lead and a piece the same size but weighing 6 lb is called code 6 sheet lead.

See if you can think of any examples in your own trade.

When working with areas, two measurements are required: **length** and **width**. It is not important which is given first, although it is good practice to think of them, and write them, in that order. You must, however, always state the units of the measurements you are using. Most measurements in the building trade are made in millimetres or metres.

Example

Let's look at something we are all familiar with: an internal door. Doors can be obtained in a variety of sizes, the most popular being 1,980 mm × 760 mm, and to find the area we multiply the length (L) by the width (W). This is written as $L \times W$ = Area, i.e.

1,980 mm × 760 mm = 1,504,800 mm^2

It will be easier, and the answer will make a lot more sense, if we first convert the measurements to metres. (If you have a problem doing this, refer back to Chapter 5.)

1,980 mm = 1.98 m; 760 mm = 0.76 m
1.98 m × 0.76 m = 1.5048 m^2

Rounded off to two significant figures $= 1.5$ m^2. It is important to
contain within the answer the units of measurement that have been used.
To simply give the answer as 1.5 will lead to confusion. Notice also that
each of the numbers being multiplied had a unit of measurement attached
to it, in this case metres. In the last chapter we learnt that a figure to be
multiplied by itself can be represented by writing a small 2 above it, i.e.
$8 \times 8 = 8^2$. It follows, therefore, that as we have multiplied metres by
metres this should also be represented with a small 2 above, i.e.
metres $\times$ metres $=$ metres^2, which can be shortened to m^2. The 'm' will
show that the units in use are metres, and the small 2 will show that it is
a measurement of area. (If you have a problem with squares or indices,
refer back to Chapter 6).

For practical purposes we can say that the surface area of the average
internal door is 1.5 m^2.

Example

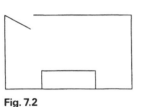

Fig. 7.2

This time we shall work out the floor area of a room as shown in the plan view in
Figure 7.2. The room measures 4.68 m by 3.25 m. Along part of one wall is a fireplace,
the base of which measures 2.60 m by 750 mm. This should not be included in the floor
area.

Floor area of room $= 4.68$ m $\times$ 3.25 m $= 15.21$ m^2
Base of fireplace $= 2.60$ m $\times$ 0.75 m $= 1.95$ m^2

As we do not want the fireplace to be included in the final floor area, we must subtract it:

15.21 m^2 $- 1.95$ m^2 $= 13.26$ m^2

If we were asked to price for sheet flooring or carpet for the room it would
be priced at so much a square metre, and by multiplying the cost of one
square metre by the area of the room a total cost can be found. For
example, if carpet cost £10 a square metre, then 13.26×10 will give a cost
of £132.60 for the carpet. (*Note*: An allowance for cutting and fitting would
have to be added to this.)

Often areas can be broken down into a series of squares and rectangles,
though at first glance they may seem to be odd shapes.

Look at the floor plan in Figure 7.3. It has been broken down into three
rectangles

$A = 1.5$ m $\times 1$ m
$B = 3.5$ m $\times 2$ m
$C = 2$ m $\times 0.75$ m

These areas worked out and added together will give the total floor area for
the room, or

Area of the floor $= A + B + C$

An alternative method of finding the area is to complete the shape into a rectangle, as shown in Figure 7.4, and from the area of the rectangle subtract the areas *a*, *b* and *c*, or:

Area of the floor = Complete rectangle − (*a* + *b* + *c*)

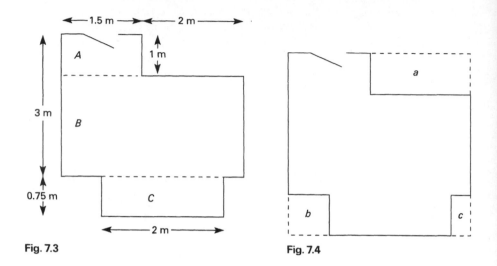

Fig. 7.3   Fig. 7.4

The method you use depends mainly on the shape and complexity of the area to be found. Often the area will not fall conveniently into rectangles, but with practice these shapes and their areas can be recognised and worked out.

Exercise 7.2

(a) What is the floor area of the room in Figure 7.3?
(b) If sheet flooring is £5.50 a square metre, what will be the cost of the flooring for the room?
(c) What is the length of the perimeter of the room?

7.3 Triangles

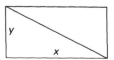

Fig. 7.5

Triangles are areas that are enclosed by three sides, and on occasions we have to determine the area of a triangle. (Triangles are looked at in more depth in Chapter 10.)

Look at Figure 7.5: the line drawn from corner to corner in the rectangle has divided it equally into two triangles. The area of one triangle is exactly half the area of the rectangle. This is true for any triangle that has the same base and perpendicular height as a rectangle, i.e. the area of a triangle with a base of x units and a perpendicular height of y units will be half the area of a rectangle with a length (base) of x units and a perpendicular height (width) of y units.

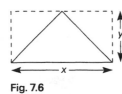

Fig. 7.6

This can be written as either

$$\text{Area of a triangle} = \frac{1}{2}(Base \times Perpendicular\ height) \text{ or}$$

$$= \frac{Base \times Perpendicular\ height}{2} \text{ or}$$

$$= (Base \times Perpendicular\ height) \div 2$$

All will give the same result. This theory is true for any type of triangle, e.g. Figure 7.5 or 7.6.

Example

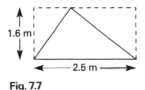

Fig. 7.7

Find the area of the triangle in Figure 7.7. It has a base of 2.5 m and a height of 1.6 m.

$$\frac{2.5 \times 1.6}{2} = \frac{4}{2} = 2 \text{ m}^2$$

The dotted rectangle enclosing the triangle has an area of 1.6 m × 2.5 m = 4 m², exactly twice that of the area of the triangle.

Exercise 7.3

You should be able to do the first question without a calculator. Try it.

(a) Find the area of a triangle that has a base of 2 m and a height 2 m.
(b) Find the area of a triangle that has a base of 9 m and a height 6 m.
(c) Find the area of a triangle that has a base of 3.9 m and a height 4.6 m.

7.4 ## Circles

Before we look at the mathematics involved in finding the areas of circles, we need first to look at some of the names given to the various parts of the circle. On a sheet of paper, draw a circle and write the names of the parts as I describe them. You may need the help of your teacher with some of them. I have drawn them at the end of this chapter (in the summary, see Figure 7.8), but you will learn better if you do them yourself.

- **Centre** The centre point is the middle of the circle. The point of your compass was positioned at this centre point and the line forming your circle was drawn at a constant distance from it.

- **Circumference** The line that is drawn around the centre point is called the circumference. The circumference of a circle is a linear measurement. If we were to stretch a piece of string around the perimeter of a car tyre the string would represent the circumference of the tyre and could be measured.
- **Radius** The distance between any point on the circumference and the centre point is called the radius.
- **Diameter** The measurement from one side of the circle to the other side, passing on its way through the centre point, is called the diameter and should be thought of as the width of a circle. The line forms two semi-circles, one each side of the diameter. The diameter is always twice the length of the radius.
- **Degree** The amount that it was necessary to turn your compass in order to draw the circumference is one *revolution*. This is split up into 360 equal divisions, known as *degrees*. The sign for degree is °. A degree can be split into smaller units called *minutes* and *seconds*. There are 60 minutes in a degree and 60 seconds in a minute.
- **Sector** A sector is an area of part of a circle, formed by two lines radiating from the centre point out to the circumference, and the portion of the circumference between the two lines.
- **Arc** An arc is that part of the circumference that acts as a boundary to the sector. The arc is a linear distance.
- **Chord** A chord is a straight line joining two points on the circumference but not passing through the centre point. The smaller area of the circle cut off by the chord is called a *minor segment*, while the remainder of the circle is called a *major segment*.

7.5 Finding the circumference or perimeter of a circle

The sixteenth letter of the Greek alphabet is pi, often written as π. Mathematicians use it as shorthand to represent the ratio of the circumference of a circle to its diameter. This kind of shorthand, as we said in Chapter 2, is often used in mathematics. Where symbols or letters are used to represent equations, numbers or functions, you do it often without thinking. Very few people write the words addition or subtraction when doing a calculation; we simple use the signs $+$ or $-$.

Pi has a numerical value of approximately $3\frac{1}{7}$ or 3.142...; the dots are another type of shorthand for saying the number has no end.

How accurate you require your answer to be when you use pi depends largely on what you are calculating. The space programme uses computers that calculate and use the number pi up to 2 million decimal places. We, however, shall limit ourselves to three decimal places, or 3.142, during this book.

To find the circumference of a circle, measure the diameter of the circle and multiply by pi. You can prove this to yourself, either in the workshop or at home. Measure the diameter of a piece of pipe, a cup or even a saucepan (the larger the diameter the better), multiply the measurement you have by 3.142, and check your answer by measuring the circumference with a flexible tape. If the two answers differ by a few millimetres it is the fault of the tape not the equation.

If the circumference of a circle can be found by multiplying the diameter by π, it follows that the circumference of a circle can also be found using the radius. Remember that the radius is half the diameter, so we shall be required to multiply by 2, which gives the formula for finding the circumference of a circle as

$$2 \times \pi \times \text{Radius} \quad \text{or} \quad 2\pi r$$

Pi is used to solve many mathematical problems found in the building industry. We can work through the following example together, making sure you understand what has been done, then you can try the exercises.

Example

A plumber has been asked to make a lead slate to fit around a 150 mm flue pipe. What length of sheet lead will be required to fit around the pipe?

If the pipe has a diameter of 150 mm, then the radius will be 75 mm. Using the formula

$$2\pi r = 2 \times 3.142 \times 75 = 471.3$$

The sheet lead will need to be 471.3 mm long. (*Note*: The plumber would also make an allowance for dressing and jointing.)

 Exercise 7.4

(a) Find the circumference of a circle that has a radius of 1 m.
(b) Find the circumference of a circle that has a radius of 300 mm.
(c) Find the circumference of a circle that has a radius of 8.9 m.

7.6 Area of a circle

The formula for finding the area of a circle is pi multiplied by the radius, the result being multiplied by the radius again, i.e.

$$\text{Area of a circle} = \pi \times \text{Radius} \times \text{Radius}$$

or, better still,

$$\text{Area of a circle} = \pi r^2$$

Remember when you square something you multiply it by itself. Therefore, radius $\times$ radius $=$ radius^2 or r^2.

Let's work out an example together, then try the exercises. Find the area of a circle whose radius is 7 metres.

Formula to be used is πr^2:

$$3.142 \times 7\,\text{m} \times 7\,\text{m} = 3.142 \times 49 = 153.958\,\text{m}^2$$

Notice that the answer is given in metres squared. This is necessary as it describes an area.

 ### Exercise 7.5

Find the areas of the circles in Exercise 7.4.

7.7 ## Measuring volumes

When calculating **volumes** three measurements are needed: length, width and depth or height. Each of the measurements is multiplied together. The formula is:

Volume = Length × Width × Depth

The unit for volume is the cubic metre and is written m^3. When you see a result that is a cubic measurement it always has three elements, length, width and depth. The formula is sometimes abbreviated to $L \times W \times D$.

Another term used in place of volume is **cubic capacity**. Capacity is associated more usually with how much a container will hold.

Example

A concrete drive is to be constructed to provide access to a garage built on the side of a house. The drive measurements are as follows: the length of the drive is 4.6 m; it is 2.9 m wide; and has a depth of 150 mm. What is the capacity of this measurement, or, in other words, how much concrete will be required to lay the drive?

The first task is to convert 150 mm to metres.

150 divided by 1,000 = 0.150

So 150 mm = 0.15 m. (Look back to Chapter 5 if you had trouble following that.)

It is just a matter of multiplication now, remembering that $L \times W \times D$ = volume. We shall take it stage by stage.

Length × Width = 4.6 m × 2.9 m = 13.34 m²

(Notice the first multiplication gives us the surface area of the drive and is identified as m².) The next task is to multiply by the depth:

13.34 m² × 0.15 m = 2.001 m³

For practical purposes, 2 m³ of concrete is required.

Let's try another example together.

Example

Find the volume of a cold water storage cistern with a base 1.5 m × 0.9 m and a height of 0.95 m. (Remember that height and depth mean the same thing for our purpose.)

Again the formula is $L \times W \times D$. Substituting the figures we get

$1.5 \times 0.9 \times 0.95 = 1.2825$ m^3

The answer we arrived at is fine if we intend to fill the cistern with concrete, sand or anything else we buy in cubic metres, but often when dealing with volumes we require more than the capacity. For instance, a plumber will want to know how much water the cistern contains and the cistern's mass or weight. So let's look briefly at a few of the physical characteristics of water.

A cubic metre of water contains 1,000 litres. A litre of water has a mass of 1 kg. Therefore, 1,000 litres of water has a mass of 1,000 kg, or one metric tonne.

We know that 1 m^3 of water contains 1,000 litres and has a mass of 1,000 kg; we also know the volume of our cistern in cubic metres. To find how many litres of water and the mass of the water in the cistern we multiply the volume of the cistern by 1,000.

Water held in cistern = 1.2825 × 1,000 = 1282.5 litres

and as 1 litre of water has a mass of 1 kg,

Mass of water held in cistern = 1282.5 kg   or   1.2825 tonnes

---

7.8 ## Volume of a cylinder or pipe

We have just learnt how to find the volume of straight-sided figures such as cisterns and concrete slabs. The formula used was $L \times W \times D =$ Volume. If we take part of that formula, it can be said that two of the measurements will provide an area. The area (length × width) will be measured as m^2, and when that is multiplied by the depth or the height we find the volume, m^3. We also looked at how to find the area of a circle, and learnt that $\pi r^2 =$ area of a circle. If that area is now multiplied by depth or height as was done for straight-sided figures, the result will be the volume of a cylinder.

Therefore, the formula to find the volume of a cylinder is

$\pi \times$ Radius $\times$ Radius $\times$ Height

This is more usually written as $\pi r^2 h$. Let's work an example out together.

Example

Find the volume of a cylinder that has a radius of 3.5 m and is 5 m high.

Using $\pi r^2 h$ substitute for the values given.

$\pi \times r \times r \times h$ becomes 3.142 × 3.5 × 3.5 × 5

which gives the cylinder as having a volume of 192.448 m^3.

And another

Example

How much water could be contained in a cylinder that has a diameter of 800 mm and a height of 1.5 m?

$3.142 \times 0.4 \times 0.4 \times 1.5 = 0.754 \text{ m}^3$
$0.754 \text{ m}^3 \times 1,000 = 754 \text{ litres}$

The first thing to notice is that the question gave the diameter of the cylinder as 800 mm. This had to be halved to obtain the radius of 400 mm or 0.4 m. Secondly, I multiplied my answer in cubic metres by 1,000 to convert it to litres.

 ### Exercise 7.5

(a) How much concrete would be required to cast a beam measuring 6 m × 0.5 m × 0.83 m?
(b) What would be the volume of a hot water cylinder which had a diameter of 600 mm and a height of 950 mm?
(c) What would be the mass of the water in (b)? Give your answer to the nearest kilogram.

Answers to exercises

7.1 (a) 7.5 m + 7.5 m + 5.5 m + 5.5 m = 26 metres of skirting
(b) 26 m at £2.10 a metre = 26 × £2.10 = cost of skirting, £54.60
(c) Cost of fitting skirting = 26 m at 90p a metre = £23.40
Cost of skirting   =   + £54.60
Total cost   =   £78.00

7.2 (a) Area of floor = $A + B + C$
area of A = 1.5 m × 1 m = 1.5 m²
area of B = 3.5 m × 2 m = 7 m²
area of C = 2 m × 0.75 m = 1.5 m²
Area of floor = 1.5 m² + 7 m² + 1.5 m² = 10 m²
(b) 10 m² at £5.50 a square metre = 10 × £5.50 = cost of sheet flooring, £55.00.
(c) Add the lengths of each wall to find the perimeter

3 m + 1.5 m + 1 m + 2 m + 2 m + 0.5 m + 0.75 m +
2 m + 0.75 m + 1 m = 14.5 m

7.3   Using the method $\dfrac{\text{Base} \times \text{Perpendicular height}}{2}$ gives

(a) $\dfrac{2 \times 2}{2} = 2 \text{ m}^2$

(b) $\dfrac{9 \times 6}{2} = 27 \text{ m}^2$

(c) $\dfrac{3.9 \times 4.6}{2} = 8.97 \text{ m}^2$

7.4   Using the formula $2\pi r$ gives

(a) $2 \times 3.142 \times 1 \text{ m} = 6.284 \text{ m}$
(b) $2 \times 3.142 \times 0.3 \text{ m} = 1.885 \text{ m}$
(c) $2 \times 3.142 \times 8.9 \text{ m} = 55.928 \text{ m}$

7.5   Using the formula πr^2 gives

(a) $3.142 \times 1 \text{ m} \times 1 \text{ m} = 3.142 \text{ m}^2$
(b) $3.142 \times 0.3 \text{ m} \times 0.3 \text{ m} = 0.283 \text{ m}^2$
(c) $3.142 \times 8.9 \text{ m} \times 8.9 \text{ m} = 248.878 \text{ m}^2$

7.6   (a)   Using the formula, Volume $= L \times W \times D$:

$6 \text{ m} \times 0.5 \text{ m} \times 0.83 \text{ m} = 2.49 \text{ m}^3$ of concrete

(b)   Using the formula πr^2 gives

$3.142 \times 0.3 \text{ m} \times 0.3 \text{ m} \times 0.950 \text{ m} = 0.269 \text{ m}^3$

(c) 1 m^3 of water has a mass of 1,000 kg

$0.269 \times 1,000 = 269 \text{ kg}$

Summary

We have been looking at:

- Measurements as used in perimeters, areas and volumes.
- The formulae for finding the areas of squares, rectangles, triangles and circles.
- The formulae for finding the volumes of straight-sided figures and cylinders.

Things to remember

- The perimeter of most two-dimensional shapes can be found by adding together the lengths of their boundaries or sides.
- Circumference of a circle $= 2 \times \pi \times \text{Radius}$,   or $2\pi r$.

- Area requires two measurements: length and width. These are multiplied together to give an answer. The units will be m^2, mm^2, etc.
- Area of a triangle = Base × Vertical height divided by 2.
- Area of a circle = π × Radius × Radius   or πr^2.
- Volume must have three measurements: length, width and depth or height. These are multiplied together to give an answer. The units will be m^3, mm^3, etc.
- Volume of a cylinder = πr^2 × height of cylinder, more usually written as $\pi r^2 h$.

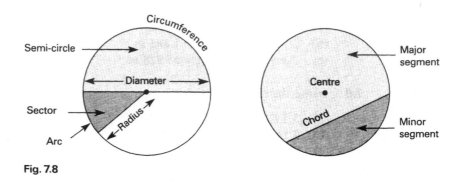

Fig. 7.8

Take a break from your studies before testing your skills on the following self-assessment questions.

Self-assessment questions

Below there are twelve questions. Take your time in answering them.

- These are not meant as a test; the questions are simply to help you learn.
- Look back at you own notes and Chapter 7 if you need help.
- Answers and comments follow the questions, but you should look at them only when you have finished or are really stuck.

SAQ 7.1   A building plot that measures 12 m by 42 m is to have a chain-link guard fence erected around its boundary, in order that it may be secured at night. Chain-link fencing is £20 a metre to supply and erect. How much fencing will need to be ordered and how much will it cost?

SAQ 7.2   A room measuring 6.40 m by 4.10 m is to have the skirting renewed. How much will be ordered from the timber merchant, and if skirting is £2.80 per metre how much will it cost? (*Note*: Do not make any allowance for door openings or wastage.)

SAQ 7.3   A painter has been asked to completely repaint a ceiling with two coats of emulsion. One litre of emulsion paint will cover an area of 12 m^2. The ceiling is 9.6 m × 5 m. How many litres of paint should be ordered?

SAQ 7.4   Calculate the floor area of the room shown in Figure 7.9.

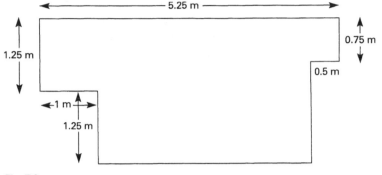

Fig. 7.9

SAQ 7.5   A garden wall 1.7 m high and 9.5 m long is to be built. If there are 60 bricks to the square metre, how many will be needed for the wall?

SAQ 7.6   A bricklayer has been asked to repoint the gable wall of a bungalow. His agreed price of £10 a square metre includes materials and raking out the joints to a depth of 20 mm. The gable is 11 m across and rises 5 m to the apex of the roof. Calculate how much money the bricklayer will receive.

SAQ 7.7   Find the length of curved skirting required for a semi-circular bay window, the diameter of which is 2.9 m.

SAQ 7.8   How much concrete will be required to make ten concrete lintels, all measuring 1.48 m by 225 mm by 150 mm?

SAQ 7.9   The air in an internal bathroom measuring 2.9 m by 3.2 m by 2.3 m high has to be changed every 15 minutes. Extraction units are measured by the amount of air they can remove per hour. What size will the extraction unit need to be? (Your answer will be in cubic metres per hour.)

SAQ 7.10   How much water could be stored in a cistern that has a base 750 mm square and a depth of 600 mm? What would be the mass of the water?

SAQ 7.11   Ten circular concrete columns are to be constructed to support a concrete floor. The height of each column is 2.74 m and the diameter 0.300 m. How much concrete is required to fill all the columns?

SAQ 7.12   What would be the mass of the water in a cylinder with a diameter of 700 mm and a height of 1.5 m?

Answers and comments

SAQ 7.1

The plot measures 12 m by 42 m and is a rectangle. To find the perimeter, the four sides are added together. The answer will be in metres. To find the cost, the answer is multiplied by the cost of a metre of fencing at £20.

12 + 12 + 42 + 42 = 108 m of chain-link fencing will need to be ordered. It will cost 108 × £20 = £2,160 to supply and erect.

(*Note*: The 108 metres represents the perimeter or boundary of the building plot.)

SAQ 7.2

The room measures 6.40 m by 4.10 m. Like the previous problem, the area is a rectangle, and we are trying to find its perimeter. The answer again will be in metres. To find the cost of the skirting the answer must be multiplied by the cost of one metre (£2.80).

6.40 + 6.40 + 4.10 + 4.10 = 21.00 m of skirting required. It will cost 21.00 × £2.80 = £58.80

(*Note*: The carpenter will need to add to this for cutting and wastage.)

SAQ 7.3

This problem is asking you to find the area of a ceiling and the amount of paint required for two coats. To find the area, the length is multiplied by the width, and the answer is divided by the area that 1 litre of paint will cover.

Area of ceiling = 9.6 m × 5 m = 48 m^2
Paint required = 48 m^2 ÷ 12 = 4 litres

As the ceiling requires two coats:

Paint required = 4 × 2 = 8 litres

SAQ 7.4

If you look carefully at Figure 7.9 you will notice that, although at first glance it appears to be a difficult shape to find the area of, it can be broken down into three rectangles, as in Figure 7.10.

The areas of the three rectangles are found by multiplying their lengths by their widths, and by adding these answers together the total area is found.

1.25 m × 1 m = 1.25 m^2
3.75 m × 2.5 m = 9.375 m^2
0.75 m × 0.5 m = 0.375 m^2
1.25 m^2 + 9.375 m^2 + 0.375 m^2 = 11 m^2

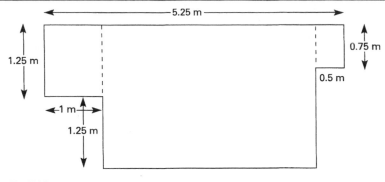

Fig. 7.10

(*Note*: Most areas found in the building trade can be broken into more convenient shapes. There are often several different methods of doing this.)

SAQ 7.5

The height and length of the garden wall are first multiplied together giving the area of the wall, the result being multiplied by 60 (the number of bricks used in a square metre of brickwork).

$$1.7 \text{ m} \times 9.5 \text{ m} = 16.15 \text{ m}^2 \times 60 = 969 \text{ bricks required.}$$

(*Note*: The bricklayer will have to allow for wastage and cutting and would probably order 1,000 bricks.)

SAQ 7.6

A gable wall on a building is a triangular area of brickwork that one end of a pitched roof bears on. The formula to find the area of a triangle is

$$\frac{\text{Base} \times \text{Perpendicular height}}{2}$$

Therefore our calculation should look like this:

$$\frac{11 \times 5}{2} \times 310 = 27.5 \text{ m}^2 \times £10 = £275$$

SAQ 7.7

The bay window is half of a circle with a diameter of 2.9 m. Our calculation then requires us to find the circumference of the circle and either divide the answer by 2 or multiply by 0.5, as both will give the same result. The formula for finding the circumference of a circle is $2\pi r$. Remembering that the radius is half the diameter, our calculation should look like this:

$$2 \times 3.142 \times 1.45 \times 0.5 = 4.556 \text{ m (to three decimal places)}$$

(*Note*: This is again an exact figure and would need to have an allowance for cutting added to it.)

SAQ 7.8

The calculation will be easier to work out if we convert all the measurements to metres. Once the area of one lintel is found it can be multiplied by 10 (the total number required) and an answer arrived at. Our calculation then should look like this:

$$1.48 \text{ m} \times 0.225 \text{ m} \times 0.150 \text{ m} = 0.04995 \text{ m}^3$$

is the amount of concrete required for one lintel. Ten lintels would require

$$0.04995 \text{ m}^3 \times 10 = 0.4995 \text{ m}^3$$

(*Note*: For practicable purposes 0.5 m³ of concrete is required.)

SAQ 7.9

The bathroom measures 2.9 m by 3.2 m by 2.3 m. To find the volume of air in the room, or the amount to be removed every 15 minutes, these are multiplied together:

$$2.9 \text{ m} \times 3.3 \text{ m} \times 2.3 \text{ m} = 21.344 \text{ m}^3$$

To find the size of the extraction unit the volume should be multiplied by 4 as the air needs to be changed every 15 minutes or 4 times an hour.

$$21.344 \text{ m}^3 \times 4 = 85.376 \text{ m}^3$$

An extraction unit capable of removing 85.376 m³ an hour is required.

SAQ 7.10

The volume of the cistern can be found by multiplying the length by the breadth by the depth (as we did in the answer to the previous problem to find the volume of the bathroom):

$$0.75 \text{ m} \times 0.75 \text{ m} \times 0.6 \text{ m} = 0.338 \text{ m}^3$$

We know that a cubic metre of water has a mass of 1,000 kg (or 1 tonne) and contains 1,000 litres of water; therefore, to find the mass and how many litres of water the cistern holds the volume, in cubic metres, must be multiplied by 1,000:

$$0.338 \text{ m}^3 \times 1,000 = 338 \text{ litres}$$

Therefore, the cistern contains 338 litres of water and has a mass of 338 kg or 0.338 tonnes.

SAQ 7.11

The formula for finding the volume of a cylinder is $\pi r^2 h$. The first thing to notice is that the problem gives us the diameter of the columns and not the radius, which the formula requires. We can solve this by remembering that the radius is half of the diameter. If the columns have a diameter of 0.300 m, then the radius will be 0.150 m. Our calculation then should look like this:

$$3.142 \times 0.150 \text{ m} \times 0.150 \text{ m} \times 2.74 \text{ m} \times 10 = 1.937 \text{ m}^3$$

(*Note*: The 10 in the calculation is the number of columns.)

SAQ 7.12

The calculation should look like this:

$$3.142 \times 0.350 \text{ m} \times 0.350 \text{ m} \times 1.5 \text{ m} \times 1{,}000 = 577 \text{ kg}.$$

CHAPTER **8** Percentages, interest, ratios and scales

Prior Knowledge If you feel you are already skilled in the subjects mentioned above, turn to page 95 and try working through the SAQs. If you find difficulty in answering them, work through this chapter.

8.1 Percentages

Percentages are often used in the building trade as a means of expressing, among other things, the amount of trade discount on offer at builders' merchants, VAT charges, wage rises, bonuses and allowances for cutting and wastage of materials, etc. The words *per cent* are Latin for 'for every hundred', therefore a **percentage** can be thought of as a fraction, with a denominator of 100. The symbol '%' is used to indicate per cent.

8.2 Converting fractions to percentages

The rule for converting fractions to percentages is: *divide the numerator of the fraction by its denominator, multiply the answer by 100 and add a % sign.*

Examples

1 Express $\frac{35}{100}$ as a percentage.

$35 \div 100 = 0.35$ Now multiply by 100.
$0.35 \times 100 = 35$ Add a % sign.
35%

2 Express $\frac{3}{8}$ as a percentage.

$3 \div 8 = 0.375$ Now multiply by 100.

$0.375 \times 100 = 37.5$ Add a % sign.

37.5%

3 Express $\frac{9}{20}$ as a percentage.

$9 \div 20 = 0.45$

We should now multiply by 100 to obtain the percentage, but first think back first to Chapter 6. There we learnt that to multiply by 10 we moved the number one place to the left and to multiply by 100 we moved the number two places to the left. So

0.45 becomes 45   Add a % sign.

45%

4 Express $\frac{7}{8}$ as a percentage.

$7 \div 8 = 0.875$ Move the numbers two places to the left.

87.5% Add a % sign.

If a mixed number (a whole number and a proper fraction) needs to be expressed as a percentage, convert it first to an improper fraction, then proceed as above.

Example

Express $2\frac{3}{4}$ as a percentage.

$= \frac{11}{4}$ If you have problems with this turn back now to Chapter 4.

$11 \div 4 = 2.75$ Move the number two places to the left.

275% Add a % sign.

 Exercise 8.1

Express the following fractions as percentages:

(a) $\frac{47}{100}$

(b) $\frac{4}{5}$

(c) $\frac{9}{10}$

(d) $\frac{5}{6}$

(e) $1\frac{4}{9}$

(f) $2\frac{5}{8}$

8.3   Converting percentages to fractions

The rule for converting percentages to fractions is: *remove the % sign, multiply by $\frac{1}{100}$ and, if possible, reduce to lowest term.*

Examples

1 Express 28% as a fraction.

$28 \times \frac{1}{100}$ Remove the % sign and multiply by $\frac{1}{100}$.

$= \frac{28}{100}$ Reduce to lowest term.

$= \frac{7}{25}$

2 Express 175% as a fraction.

$175 \times \frac{1}{100}$ Remove the % sign and multiply by $\frac{1}{100}$.

$= \frac{175}{100}$ Reduce to lowest term.

$= \frac{7}{4}$ Convert to a mixed number.

$= 1\frac{3}{4}$

3 Express 22.4% as a fraction.

$22.4 \times \frac{1}{100}$ Remove the % sign and multiply by $\frac{1}{100}$.

$= \frac{22.4}{100}$ To remove the decimal point we must multiply the numerator and denominator by 10 or move the numbers one place to the left, adding 0 if necessary.

$= \frac{224}{1000}$

$= \frac{56}{250}$ Reduce to lowest term.

$= \frac{28}{125}$

Exercise 8.2

Rewrite the following percentages as fractions:

(a) 20%
(b) 45%
(c) 12.5%
(d) 95%

8.4 ## Converting decimals to percentages

The rule for converting decimals to percentages is: *multiply by one hundred or move the number two places to the left and add a % sign.*

Examples

1 Express 0.23 as a percentage.

$0.23 \times 100 = 23$ Move the number two laces to the left.

23% Add a % sign.

2 Express 0.4 as a percentage.

$0.40 \times 100 = 40$ A zero was placed after the 4 to allow the number to be moved two places to the left.

40%   Add a % sign.

Exercise 8.3

Rewrite the following as percentages:

(a) 0.394
(b) 3.42
(c) 0.048
(d) 103.2

8.5   Converting percentages to decimals

The rule for converting percentages to decimals is: *remove the % sign and multiply by $\frac{1}{100}$ by moving the number two places to the right, adding zeros if necessary.*

Example

1 Express 19% as a decimal.

$19 \times \frac{1}{100} = 0.19$ Move the numbers two places to the right.

2 Express 7.5% as a decimal.

0.075   This time I have simply moved the number. A zero has been placed before the 7 to allow the number to be moved two places to the right.

Exercise 8.4

Rewrite the following as decimals:

(a) 17%
(b) 104.3%
(c) 9.8%
(d) 0.37%

8.6   Practical applications

A builders' merchant offers a 4% discount for prompt payment of accounts. A local building firm orders materials that have a list price of £950 (without the discount). They intend to settle their account by paying cash as soon as the materials are delivered. How much will they pay?

By rewriting the 4% as either a decimal or a fraction and using it to multiply the original amount by £950, the discount can be found.

4% as a decimal $= 0.04$

£950 $\times$ 0.04 $=$ £38 discount

4% as a fraction$= \frac{4}{100} = \frac{1}{25}$

$950 \times \frac{1}{25} = \frac{950}{1} \times \frac{1}{25}$; 950 and 25 cancel out to $\frac{38}{1} \times \frac{1}{1} =$ £38 discount

list price   £950
discount $- \underline{£38}$
                   £912 to pay

Both methods give the result that the builder will get a discount of £38. This gives us the formula

$A = P \times B$

where

$A =$ amount or 'what is...?'
$P =$ percentage expressed as a decimal or a fraction
$B =$ base or total number and usually follows the word 'of'

As part of the contract the same builder has agreed to the customer retaining 12% of the total cost of the work for six months after the job is finished as a guarantee of workmanship. The total cost of the finished work was £3,240. How much will the customer retain?

What is (A) 12% (P) of £3,240 (B)?

I will work out the answer by rewriting the 12% as a decimal and leave you to work it out by changing the 12% to a fraction. Remember that both answers should be the same.

12% as a decimal $= 0.12$
£3,240 $\times$ 0.12 $=$ £388.80   Retention.
  (B) $\times$ (P) $=$ (A) Remember $B \times P$ is the same as $P \times B$.

Whether you change the percentage to a fraction or a decimal is not important; the method you use should be the one you are most comfortable with. As you are probably using a calculator to help you, I shall do the rest of the examples by converting the percentages to decimals.

The customer has decided to retain £400; what percentage of £3,240 is this or what percentage of £3,240 is £400?

The formula $A = P \times B$ can be rearranged to $P = \dfrac{A}{B}$ to find the answer.

Percentage $= \frac{400}{3240}$ or $(400 \div 3,240) =$

0.1234567 $\times$ 100   or move the decimal point two places to the right to
                            convert to a percentage.

12.346%   Correct to three decimal places.

The formula $A = P \times B$ can also be rearranged to answer problems such as:

27 is 12% of what number?

This time we need to find B and can rearrange the formula as $B = \dfrac{A}{P}$. The 12% must of course still be expressed as a decimal.

$B = \frac{27}{0.12} = 225$ 27 is 12% of 225

As you can see, by using or rearranging the formula $A = P \times B$ most problems relating to percentages can be solved. (*Note*: The rearranging of formulae is dealt with in more depth in Chapter 11.)

Look carefully at the three examples below. The same figures have been used each time. Try to relate the form of the question to the formula used, then do Exercise 8.5.

Examples

1   What is 30% of 80?   Formula to be used $A = P \times B$
    $0.3 \times 80 = 24$

2   What percentage of 80 is 24?   Formula to be used $P = \dfrac{A}{B}$
    $\frac{24}{80} = 24 \div 80 = 0.3 = 30\%$

3   24 is 30% of what number?   Formula to be used $B = \dfrac{A}{P}$
    $\frac{24}{0.3} = 24 \div 0.3 = 80$

Exercise 8.5

(a)   What is 27.5% of 100?
(b)   50 is 25% of what number?
(c)   24 is what percentage of 80?
(d)   What percentage of £180 is £25?
(e)   £16 is 32% of how much?
(f)   What is 12% of £104?

8.7   # Percentage increases, decreases and discounts

Often we are told something has increased, and on rare occasions has decreased, in price. The amount of **increase** or **decrease** is usually given as a percentage of the original price and it is left to us to work out the new cost.

The amount by which a quantity is increased or decreased can be found by multiplying the original quantity by the percentage of increase or decrease expressed as a decimal, as we did earlier, and by adding this figure to or subtracting it from the original quantity the new cost can be found, i.e.

Amount of increase = New price − Old price
Amount of decrease = Old price − New price

Example

Due to a shortage in raw material the price of 28 mm copper tube has increased by 20%. Before the increase 28 mm copper tube sold for £2.90 per metre. What price should you now expect to pay per metre for 28 mm copper tube?

Using the formula $A = P \times B$ gives

$0.2 \times £2.90 = 0.58p$ 28 mm copper tube has increased in price by 58p per metre. By adding this to the original cost the new price is found.

£2.90
£0.58
—————
£3.48   The new price of 28 mm copper tube.

In the unlikely event that the price of copper tube had decreased by 20%, or the merchant was offering a discount of 20%, the 58p would have been subtracted from the original price.

Example

A builders' merchant is selling cement with a discount of 22% for bulk purchase. Cement usually retails at £4.50 a bag. What would be the new price per bag if a bulk order was placed?

$0.22 \times £4.50 = 0.99p$ A reduction of 99p per bag is available for bulk purchase. By subtracting this from the original cost the new price is found.

£4.50
£0.99
—————
£3.51   The bulk purchase price per bag of cement.

Exercise 8.6

Add 12% to the following:

(a) £14.25
(b) £1,300
(c) £43.50

Discount 8% from the following:

(d) £3,050
(e) £68.40
(f) £1.50

The percentage increase or decrease can be determined, providing the new cost and the old cost are known, by using the formula

$$\text{Percentage} = \frac{\text{Amount increase or decrease}}{\text{Original amount}} \times 100$$

Example

A spare part for a boiler last year cost £27.40 and this year costs £29.80. What percentage increase does this represent?

We first need to determine the amount of the increase by subtracting the old price from the new price:

$$\begin{array}{r} £29.80 \\ -£27.40 \\ \hline £2.40 \end{array} \qquad \text{Amount of increase.}$$

Percentage = 2.40 ÷ 27.40 = 0.087591 × 100
= 8.7591
= 8.759%   Correct to three decimal places.

Note: Remember a quick way of multiplying by 100 is to move the number two places to the left.

Example

A new van sold for £13,500. Two years later, its value was assessed as £8,300. What was the percentage decrease in value over the two-year period?

Determine the amount of the decrease by subtracting the price of the van when new from the price it is worth now.

$$\begin{array}{r} £13,500 \\ -\ \ £8,300 \\ \hline £5,200 \end{array} \qquad \text{Amount of decrease.}$$

Percentage = 5,200 ÷ 13,500 = 0.3851851   Move the number two place to the
= 38.51851%   left to express as a percentage.
= 38.519%   Correct to three decimal places.

Exercise 8.7

Fill in the percentage increase or decrease for the following, giving your answer correct to one decimal place:

	Old price	New price	Percentage
(a)	£14.56	£17.48	
(b)	£18.45	£28.95	
(c)	£127.56	£92.74	
(d)	£4,059	£3,426	

8.8   # Simple and compound interest

If you borrow money for a period, it is very likely you will be charged for doing so. If the amount you borrow and the stated rate of **interest** remain the same, the interest paid every year will be the same. This is known as **simple** interest.

The amount outstanding is equal to the **principal** (the amount you borrowed) plus the interest at any given time and can be worked out using the following formula:

$$\text{Interest} = \text{Principal} \times \text{Rate of interest} \times \text{Number of years}$$

(The number of years is often written as *per annum* or p.a.)

Example

If you borrowed £5,000 at an interest rate of 7% over three years, at the end of the period you would have paid

£5,000 × 0.07 × 3 = £1,050 of interest

(I converted 7% to the decimal 0.07. I could instead have multiplied by $\frac{7}{100}$, as both will give the same result.)

Added to the principal = £5,000 + £1,050 = £6,050

You may, of course, be fortunate enough to be able to deposit money in a bank rather than borrow it, in which case your £5,000 would have given you £1,050 profit (at the same rate of interest).

This form of interest assumes that, at the end of each year, you have either paid the interest you owe or removed the profit from the bank, and so at the start of each year you either owe or have on deposit the same amount.

If you choose neither to remove nor to pay off the interest at the end of the year, then the sum you have invested or borrowed will increase by that year's interest and the interest on the second year will also increase as the principal sum will have increased. This growing interest is called **compound** interest. The interest and the growth of the principal can be calculated year by year.

Example

Calculate the compound interest on £5,000 at 7% p.a. over three years.

Principal		= £5,000
Interest after 1st year	£5000 × 0.07	= £350
Principal after 1 year		= £5,350
Interest after 2nd year	£5,350 × 0.07	= £374.50
Principal after 2 years		= £5,724.50
Interest after 3rd year	£5,724.50 × 0.07	= £400.72
Principal after 3 years		= £6,125.227

The total compound interest over 3 years is £1,125.22. This type of interest is usually employed by banks and building societies.

(*Note*: There is a formula for working out compound interest. It is, however, beyond the scope of this book.)

Exercise 8.8

Calculate the compound interest on £3,000 at 5% p.a. over two years.

8.9   Ratios

A **ratio** is one method of describing one quantity in relation to another. For example, if mortar is to be made from a mixture of sand and cement in the ratio of 7 to 2, this requires that for every seven parts of sand used in the mix two parts of cement must be added, giving a total of nine parts.

The ratio 7 to 2 can be written as either $\frac{7}{2}$ or 7 : 2 and means that the '7' quantity is to be $3\frac{1}{2}$ times larger than the '2' quantity. The ratio 7 : 2 could also be written as 3.5 : 1.

Example

Suppose that a bonus of £180 is paid to a bricklayer and his hod carrier, and they decide between them that a fair method of dividing the money would be in the ratio of 7 : 2; that is, for every £7 the bricklayer gets, the hod carrier receives £2.

To find how much each man receives, the money is first divided into nine equal parts (7 + 2 = 9) with the bricklayer receiving 7 parts and the hod carrier 2:

£180 ÷ 9 = £20
20 × 7 = £140   Bricklayer's bonus.
20 × 2 = £40   Hod carrier's bonus.

The ratio can also be found if only the quantities are known; for example, the hod carrier may have felt that £40 was not a fair share and convinced the bricklayer to give him another £20. The hod carrier now receives £40 + £20 = £60 and, of course, the bricklayer's share is reduced by £20 to £120. The ratio now is $\frac{120}{60}$ or 120 : 60, which can be better expressed as $\frac{2}{1}$ or 2 : 1. Notice how, as one bonus increases, the other decreases.

There are often more than two quantities involved in a ratio, but the procedure is nevertheless the same.

Example

A painter needs 1 litre of paint to be mixed from four different colours in the ratio of 68 : 7 : 4 : 1. How much of each colour should be mixed? (Remember 1 litre = 1,000 millilitres)

68 + 7 + 4 + 1 = 80   Add together the ratios.
1,000 ml ÷ 80 = 12.5 ml   This is each part or share.

Colour *a*   68 × 12.5 ml = 850 ml
Colour *b*   7 × 12.5 ml = 87.5 ml
Colour *c*   4 × 12.5 ml = 50 ml
Colour *d*   1 × 12.5 ml = 12.5 ml

The answer can be checked by adding the quantities together

850 ml + 87.5 ml + 50 ml + 12.5 ml = 1,000 ml or 1 litre.

Exercise 8.9

(a)   30 buckets of mortar are required with a ratio of sand to cement of 4 : 1. How many buckets of sand and how many buckets of cement are required?

(b)   A bonus of £100 is to be divided between a foreman and a painter in the ratio of 3 : 2. How much will each person receive?

(c)   Concrete is made up from aggregate, sharp sand and cement in the ratio of 6 : 3 : 1. How much of each would be required to make a cubic metre of concrete?

8.10   Scales

Much of the work you are likely to be involved in, while employed in the building trade, will be carried out by working directly from prepared drawings or plans. When drawings of an object are being prepared, it is possible, if the object is not too large, to make the drawing to the actual size of the object. However, for the majority of drawings used in the building trade this is not a viable proposition and the drawings are required to be made much smaller than the object.

This reduction in size is overcome by drawing the plans to scale. The scale a plan is drawn to is given as a ratio; for example, a plan of a building drawn to a scale of 1 : 20 signifies that one unit of measurement on the drawing represents twenty of the same units on the actual building.

By multiplying the measurement taken from the drawing by the scale, the actual measurement can be found.

Example

A brick wall measuring 300 mm on a plan drawn to a scale of 1 : 20 would be

300 mm × 20 = 6,000 mm or 6 m long

Scale rules, similar to the one shown in Figure 8.1, are available for purchase at most good office supplies shops. These are used for reading and preparing scale drawings and do away with the need to multiply the scale measurement by the scale ratio each time a measurement is required.

The following are the preferred scales and their uses:

Type of drawing	Scale		
Location (street plan)	1 : 2,500	1 : 1,250	
Site plan (position of building, etc., on site)	1 : 500	1 : 200	
	1 : 100	1 : 50	
Component (position of doors, windows, etc.)	1 : 100	1 : 50	1 : 20
Detail (details of components)	1 : 10	1 : 5	1 : 1

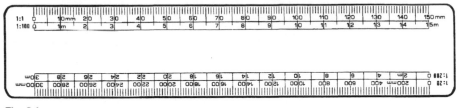

Fig. 8.1

Exercise 8.10

A line on a drawing measures 150 mm. State the length it would represent using a scale of:

(a)   1 : 1,250
(b)   1 : 200
(c)   1 : 50
(d)   1 : 5

None of the self-assessment questions relate to scales and their uses. Instead, the reader should try to obtain a scale rule and practise its use.

Answers to exercises

8.1   (a) 47%; (b) 80%; (c) 90%; (d) 83.333%; (e) 144%; (f) 262.5%

8.2   (a) $\frac{1}{5}$; (b) $\frac{9}{20}$; (c) $\frac{1}{8}$; (d) $\frac{19}{20}$

8.3   (a) 39.4%; (b) 342%; (c) 4.8%; (d) 10,320%

8.4   (a) 0.17; (b) 1.043; (c) 0.098; (d) 0.0037

8.5   (a) 27.5; (b) 200; (c) 30%; (d) 13.89%; (e) £50; (f) £12.48

8.6   (a) £15.96; (b) £1,456; (c) £48.72; (d) £2,806; (e) £62.93; (f) £1.38

8.7   (a) 20%; (b) 56.9%; (c) 27.3%; (d) 15.6%

8.8   £307.50

8.9   (a)   24 of sand, 6 of cement
          (b)   £60 and £40
          (c)   0.6 m^3 of aggregate; 0.3 m^3 of sand; 0.1 m^3 of cement

8.10   (a) 187,500 mm or 187.5 m; (b) 30,000 mm or 30 m
           (c) 7,500 mm or 7.5 m; (d) 750 mm or 0.75 m

Summary

We have been looking at:

- Percentages.
- Converting fractions to percentages.
- Converting percentages to fractions.
- Converting decimals to percentages.
- Converting percentages to decimals.
- Practical applications.
- Percentage increase, decrease and discounts.
- Simple and compound interest.
- Ratios.

Things to remember

Converting fractions to percentages

Divide the numerator of the fraction by its denominator, multiply the answer by 100 and add a % sign. If a mixed number (a whole number and a proper fraction) needs to be expressed as a percentage, convert it first to an improper fraction.

Converting percentages to fractions

Remove the % sign and multiply by $\frac{1}{100}$ and then reduce, if necessary, to lowest term.

Converting decimals to percentages

Multiply the decimal number by 100 or move the number two places to the left. Add a % sign.

Converting percentages to decimals

Remove the % sign and either multiply by $\frac{1}{100}$ or move the number two places to the right, adding zeros if necessary.

Practical applications

The formula $A = P \times B$, where

A = amount or 'what is...?'
P = percentage expressed as a decimal or a fraction
B = base or total number and usually follows the word 'of'

can be rearranged as

$$P = \frac{A}{B} \quad \text{or} \quad B = \frac{A}{P}$$

and can be used to solve most problems relating to percentages.

Percentage increases, decreases or discounts

The percentage increase, decrease or discount given can be determined, providing the new cost and the old cost are known, by using the formula

$$\text{Percentage} = \frac{\text{Amount of increase or decrease}}{\text{Original amount}} \times 100$$

Simple and compound interest

Simple interest can be found by using the formula

Interest = Principal × Rate of interest × Number of years

This form of interest assumes that, at the end of each year, you have either paid the interest you owe or removed the profit from the bank, and so at the start of each year you either owe or have on deposit the same amount.

If you choose neither to remove nor to pay off the interest at the end of the year, then the amount you have invested or borrowed will increase by that year's interest and the interest on the second year will also increase as the principal sum will have increased. This growing interest is called compound interest.

Ratios

A ratio is one method of describing one quantity in relation to another.

Scales

Scales use ratios as a method of converting measurements and dimensions on a drawing to full size.

Take a break from your studies before testing you skills on the following self-assessment questions.

Self-assessment questions

Below there are twelve questions. Take your time in answering them.

- These are not meant as a test; the questions are simply to help you learn.
- Look back at you own notes and Chapter 8 if you need help.
- Answers and comments follow the questions, but you should look at them only when you have finished or are really stuck.

SAQ 8.1   Rewrite the following fractions as percentages correct to two decimal places:

(a) $\frac{9}{16}$; (b) $\frac{13}{32}$; (c) $\frac{5}{12}$; (d) $3\frac{4}{5}$

SAQ 8.2   Rewrite the following percentages as fractions:

(a) 26%; (b) 38%; (c) 22.5%; (d) 308%

SAQ 8.3   Rewrite the following decimals as percentages:

(a) 3.907; (b) 0.046; (c) 0.0003; (d) 1.6

SAQ 8.4   Rewrite the following percentages as decimals:

(a) 16%; (b) 2%; (c) 5.9%; (d) 236.7%

SAQ 8.5   (a)   A plastering company, after estimating the cost of labour and materials for a job, adds 23% to this figure for profit and general running expenses of the company. If a job is priced at £3,642.28 for labour and materials, how much will the final estimate be?

(b)   The materials for the work above cost £840. What percentage is this of the final estimate? Give your answer to two decimal places.

SAQ 8.6   A builders' merchant has decided to increase the price of the more popular items carried in stock by 8%. Printed below is a selection of the old prices. Complete the list by putting in the new prices.

Stock no.	Old price	New price
378	£14.28	
379	£27.34	
421	£104.20	
437	£2.40	
468	£108.80	
476	£2.28	
486	£16.40	
508	£11.40	
539	£84.00	
567	£86.67	
598	£254.69	
614	£35.26	
676	£1.40	
747	£42.23	
791	£94.76	
842	£71.94	

SAQ 8.7   The same builders' merchant has also decided to offer a 5% discount on the less popular stock items. Printed below is a selection of the old prices.

Complete the list by filling in the discounted prices.

Stock no.	Old price	Discounted price
224	£39.67	
278	£14.82	
306	£10.61	
358	£31.71	
390	£79.00	
418	£50.62	
493	£92.00	
499	£31.42	
504	£95.03	
572	£44.53	
630	£84.27	
673	£34.23	
700	£56.00	
712	£49.23	
745	£11.64	
893	£38.20	

SAQ 8.8   A builder buys a plot of land for £27,300, which is later sold for £31,500. What is the percentage increase in value? Give your answer correct to two decimal places.

SAQ 8.9   If you invested £3,000 in a building society with an interest rate of 5% p.a. and left the interest to accumulate for four years, by how much would your original investment grow?

SAQ 8.10   A brick manufacturer produces common bricks and face bricks in a ratio of 5 : 2. How many of each would be produced in a firing of 280,000 bricks?

Answers and comments

SAQ 8.1

Using the method of dividing the numerator by the denominator of the fraction, multiplying the result by 100 and adding a % sign gives:

(a) $9 \div 16 = 0.5625$
$0.5625 \times 100 = 56.25\%$ (correct to two decimal places)
(b) $13 \div 32 = 0.40625$
$0.40625 \times 100 = 40.625$
$40.625 = 40.63\%$ (correct to two decimal places)
(c) $5 \div 12 = 0.4166666...$
$0.4166666 \times 100 = 41.66666...$
$41.66666... = 41.67\%$ (correct to two decimal places)

(d)   As $3\frac{5}{5}$ is a mixed number it must first be changed to an improper fraction. As you learnt in Chapter 4, this is done by multiplying the whole number by the denominator, adding the numerator and putting the result over the original denominator.

$$3 \times 5 = 15$$
$$15 + 4 = 19 = \tfrac{19}{5}$$
$$19 \div 5 = 3.8$$
$$3.8 \times 100 = 380$$
$$380\%$$

SAQ 8.2

To convert a percentage to a fraction, remove the % sign and rewrite as a fraction with the percentage as the numerator and 100 as the denominator; reduce if necessary to lowest term.

(a) $26\% = \frac{26}{100}$ 26 and 100 will both divide by 2 to give
      $\frac{13}{50}$ reduced to its lowest term.

(b) $38\% = \frac{38}{100}$ 38 and 100 will both divide by 2 to give
      $\frac{19}{50}$ reduced to its lowest term.

(c) $22.5\% = \frac{22.5}{100}$ To remove the decimal point we must multiply the numerator and denominator by 10 or move the numbers one place to the left.
      $\frac{225}{1000}$ 225 and 1,000 will both divide by 5 to give
      $\frac{45}{200}$ 45 and 200 will again both divide by 5 to give
      $\frac{9}{40}$ reduced to its lowest term.

(d) $308\% = \frac{308}{100} = 3\frac{8}{100} = 3\frac{2}{25}$

SAQ 8.3

To rewrite a decimal as a percentage, multiply by 100 or move the number two places to the left. Add a % sign.

(a) $3.907 \times 100 = 390.7$
      390.7% Add a % sign.

(b) $0.046 \times 100 = 4.6$
      4.6%

(c) $0.0003 = 0.03$ I have moved the number two places left rather
      0.03% than multiply by 100. Doing either will give the same result.

(d) $1.6 = 160$ A zero was placed after the 6 to allow the number to
      160% be moved two places to the left. Multiplying by 100 would have given the same result.

SAQ 8.4

A percentage can be changed to a decimal by removing the percentage sign and either multiplying by $\frac{1}{100}$ or moving the number two places to the right, adding zeros if necessary.

(a) $16\% = 16 \times \frac{1}{100} = 0.16$

(b) $2\% = 2 \times \frac{1}{100} = 0.02$

(c) $5.9\% = 5.9 = 0.059$ I have moved the number two places to the right rather than multiply by $\frac{1}{100}$.

(d) $236.7\% = 2.367$ I have moved the number two places to the right rather than multiply by $\frac{1}{100}$. Remember that either method will give the same result.

SAQ 8.5

(a) We first need to find 23% of £3,642.28 and add this to £3,642.28. To find 23% of £3,642.28, the 23% is rewritten as a decimal and used to multiply £3,642.28

$$23\% = 0.23$$ Written as a decimal.

$$
\begin{array}{r}
£3642.28 \\
\times \quad 0.23 \\
\hline
1092684 \\
728456 \\
\hline
837.7244
\end{array}
$$
As we are dealing with money this should be rounded off to £837.72.

$$
\begin{array}{r}
£3642.28 \\
+ \quad £837.72 \\
\hline
£4480.00
\end{array}
$$
The cost of the final estimate.

(b) The question is asking what percentage of £4,480 is £840? This can be found using the rearranged formula $P = \dfrac{A}{B}$, where P equals the percentage, A equals the amount (£4,480) and B equals the base (£840) to give

$$840 \div 4480 = 0.1875$$ Move the number two places to the left to
$$18.75\%$$ convert to a percentage.

SAQ 8.6

The amount by which each item is increased can be found by multiplying the old price by the percentage increase expressed as a decimal and by adding this figure to the old price. For example, 8% rewritten as a decimal = 0.08.

$$£14.28 \times 0.08 = £1.14$$
$$£14.28 + £1.14 = £15.42$$

An alternative and somewhat quicker method is to rewrite the percentage as a decimal and add one to it. Using this to multiply the old price with, the answer obtained will be the new price. For example 8% rewritten as a decimal $= 0.08 + 1 = 1.08$.

$$£14.28 \times 1.08 = £15.42$$

Either method will give the same result.

Stock no.	Old price	New price
378	£14.28	£15.42
379	£27.34	£29.53
421	£104.20	£112.54
437	£2.40	£2.59
468	£108.80	£117.50
476	£2.28	£2.46
486	£16.40	£17.71
508	£11.40	£12.31
539	£84.00	£90.72
567	£86.67	£93.60
598	£254.69	£275.07
614	£35.26	£38.08
676	£1.40	£1.51
747	£42.23	£45.61
791	£94.76	£102.34
842	£71.94	£77.70

SAQ 8.7

The discounted price can be found by multiplying the old price by the percentage decrease (5%) expressed as a decimal and by subtracting the result from the old price. For example, 5% rewritten as a decimal = 0.05.

$$£39.67 \times 0.05 = £1.98$$
$$\begin{array}{r} £39.67 \\ -\ £1.98 \\ \hline £37.69 \end{array}$$

Stock no.	Old price	Discounted price
224	£39.67	£37.69
278	£14.82	£14.08
306	£10.61	£10.08
358	£31.71	£30.12
390	£79.00	£75.05
418	£50.62	£48.09
493	£92.00	£87.40
499	£31.42	£29.85
504	£95.03	£90.28
572	£44.53	£42.30
630	£84.27	£80.06
673	£34.23	£32.52
700	£56.00	£53.20
712	£49.23	£46.77
745	£11.64	£11.06
893	£38.20	£36.29

SAQ 8.8

The percentage increase can be found by using the formula

$$\text{Percentage} = \frac{\text{Amount of increase}}{\text{Original amount}} \times 100$$

The amount of the increase is found by deducting the price the builder paid for the land (the original amount £27,300) from the price it was sold for (£31,500).

$$
\begin{array}{r}
£31,500 \\
-\,£27,300 \\
\hline
£4,200 \quad\quad \text{Amount of increase}
\end{array}
$$

$$\text{Percentage} = \frac{4,200}{27,300} \times 100 = 4.200 \div 27,300 = 0.1538461$$

Remember: moving the number two places to the left gives the same result as multiplying by 100:

0.1538461 × 100 = 15.38461%
15.38%   Correct to two decimal places.

SAQ 8.9

Calculate the compound interest on £3,000 at 5% p.a. over four years

Principal		= £3,000
Interest after 1st year	£3,000 × 0.05	= £150
Principal after 1 year		= £3,150
Interest after 2nd year	£3,150 × 0.05	= £157.50
Principal after 2 years		= £3,307.50
Interest after 3rd year	£3,307.50 × 0.05	= £165.38
Principal after 3 years		= £3,472.88
Interest after 4th year	£3,472.88 × 0.05	= £173.65
Principal after 4 years		= £3,646.52.

The original investment of £3,000 would have grown by £646.52.

SAQ 8.10

To find the number of common bricks and the number of facing bricks in a firing of 280,000 bricks, where the bricks are made in a ratio of 5 : 2, first add the ratios together and divide the total amount of bricks:

5 + 2 = 7
280,000 ÷ 7 = 40,000

This figure is now multiplied first by 5 to obtain the number of common bricks in the firing and then by 2 to find the number of facing bricks.

40,000 × 5 = 200,000   Common bricks
40,000 × 5 = 80,000   Facing bricks

CHAPTER 9 Lines and angles

Prior Knowledge   If you feel you are already skilled in the subjects mentioned above, turn to page 109 and try working through the SAQs. If you find difficulty in answering them, work through this chapter.

9.1 Points and lines

In **geometry** (the word *geometry* comes from the Greek language and is roughly translated as meaning 'Earth measurement') a point has no length, area or volume; in fact it has no size at all. How then can it be drawn? The one thing a point does have is position, i.e. it occupies a certain place, and that place can be marked, with a sharpened pencil. Much of geometric construction is concerned with finding the exact location of these points. Points are named or referred to by capital letters such as point A or point B.

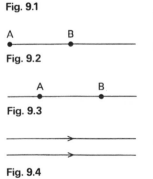

Fig. 9.1

Fig. 9.2

Fig. 9.3

Fig. 9.4

- The shortest distance between any two points is a straight line, as in Figure 9.1. This line between two **endpoints** such as A and B is a **line segment**.
- A line that stopped or started at one endpoint and passed through the other point as in Figure 9.2, is called a **half line**.
- A line passing through both points A and B, as in Figure 9.3, will stretch on to infinity – that is, it will have no end. However, as it lies on the points A and B it can still be called the line AB. The fact that many of the lines we draw in geometry do not have an end is not important. If you look at the examples on the next few pages you will see that most of the lines do not have endpoints and really could go on for ever.
- If two lines are drawn on a plane surface they either **intersect** (have a common point where they meet) or run **parallel** with each other. Parallel lines are lines that are always the same distance apart no matter how far they are extended. To show that lines are parallel we mark each of the lines with an arrow, as in Figure 9.4.

9.2 Angles

If two lines are not parallel they must cross, or **intersect**, at some point. The point at which they intersect is called the **vertex**. As the lines are going in different directions they form internal and external angles at the vertex (i.e. where they meet). The length of the lines does not affect the angles. Although two angles shown in Figure 9.5 may look different, they are both

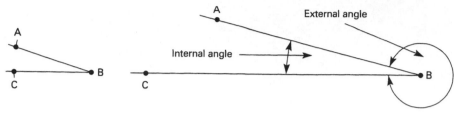

Fig. 9.5

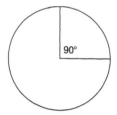

Fig. 9.6

angles of 25 degress (written as 25°). Think back to Chapter 7 and the circle you drew. There we found that you had to turn your compass one full revolution in order to draw the circumference of the circle, and that the circumference was split into 360 equal divisions, called degrees. It follows then that if a complete circle is 360°, a quarter of a circle will be $360 \div 4 = 90°$, as shown in Figure 9.6, and that 180° would be half a circle, while 270° would be three-quarters of a circle.

Exercise 9.1

Draw a circle, and mark approximately on the circumference.

(a) 90°
(b) 180°
(c) 270°
(d) 360°

9.3 Measuring and drawing angles

Protractors

What you have after completing Exercise 9.1 is a very rudimentary method of measuring angles. However, as you would expect, a commercially produced means of doing this is already available. It is called a **protractor** and it is capable of measuring many more angles.

Protractors are available in two types: **circular** incorporating 360°, and **semi-circular** measuring 0–180° (shown in Figure 9.7). They are both usually constructed of clear plastic.

Measuring angles

To measure an angle follow the steps outlined below:

1. Place the protractor over the angle such that its centre lies on the point where the two lines of the angle meet.

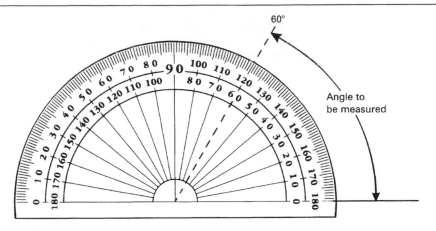

60°

Angle to
be measured

Fig. 9.7

2. Turn the protractor slowly until the line on the bottom of the protractor marked 0°–180° covers one of the lines of the angle to be measured, as in Figure 9.7.
3. Read off the size of the angle from the scale at the side of the protractor where the second line of the angle crosses it.

Note: Most protractors have two scales for ease of use. Care must be taken that the correct scale is being used.

Drawing angles

To draw an angle follow the steps outlined below:

1. Draw a straight line and make a mark 'A' where you require the angle to start.
2. Place the protractor so that the line on the bottom of the protractor marked 0°–180° lies exactly over the line you have drawn.
3. Slowly move the protractor along the line until its centre lies on your start mark 'B'.

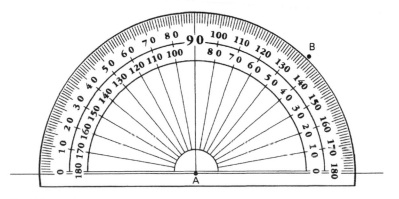

Fig. 9.8

4. Mark off the position where the protractor indicates the required angle 'B' (45° in Figure 9.8).
5. Remove the protractor and join the two marks 'A' and 'B'.

Set squares

Also available for drawing or checking angles are **set squares**. A set square is a rigid frame, two sides of which form a right angle (90°). Set squares for drawing are normally triangular in shape, made from clear plastic and are available to measure either 90°, 60° and 30° or 90°, 45° and 45°, as shown in Figure 9.9.

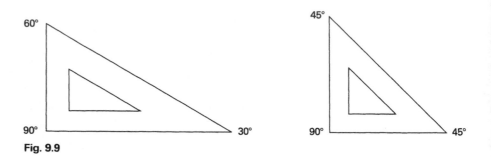

Fig. 9.9

In building work the most commonly used set squares measure 90° or 45°. Figure 9.10 shows:

(a) A carpenter's try square, used for setting out right angles and testing planed timber for square edges.
(b) An adjustable try square, which has both 90° and 45°.
(c) A builder's square, which is usually made on site from 32 mm × 100 mm timber and can be anything up to 2 metres in length.

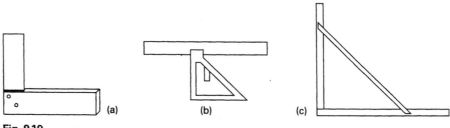

Fig. 9.10

9.4 Notation, types and names of angles

An angle in a diagram can be indicated by a small (lower case) letter and an arc drawn between the lines that form the angle, as in Figure 9.12.

Alternatively, the three points – i.e. the point at the vertex and a point on each of the two half lines – can be lettered and the angle referred to by these. The angles in Figure 9.5 would be called the angle ABC or CBA, as both mean the same angle. The letter at the vertex is always in the centre and is sometimes marked $\hat{B}$. Often a small angle $\angle$ is placed before the letters as shorthand for 'the angle'.

(a) Figure 9.11 shows a **right** angle or 90° angle; there are two right angles in 180° and four in 360°. They are often marked with a small square in the angle.

(b) The angles shown in Figure 9.12 are all **acute** angles. An acute angle is an angle that is *less* than 90°.

Fig. 9.11

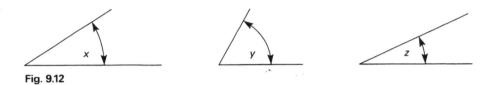

Fig. 9.12

(c) The angles shown in Figure 9.13 are all **obtuse angles**. An obtuse angle is one that is *greater* than 90° but *less* than 180°. (180° would be a straight line.)

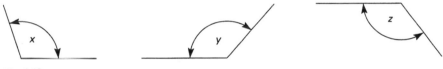

Fig. 9.13

(d) The angles in Figure 9.14 are all **reflex** angles. Reflex angles are angles that are *greater* than 180° and *less* than 360°. (360° would be a complete circle.)

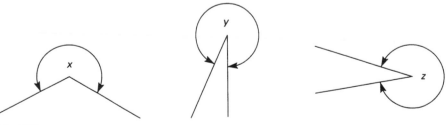

Fig. 9.14

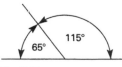

Fig. 9.15

Two angles which add up to 180° are called **supplementary** angles; that is, they supplement each other to form a straight line, as in Figure 9.15, which shows that 65° is the supplement or supplementary angle of 115°; similarly, an angle of 115° is the supplement of an angle of 65°. The sum of the angles on a straight line is 180°.

Examples

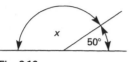

Fig. 9.16

1 Find the supplementary angle x in Figure 9.16.

$$x = 180° - 50° = 130°$$

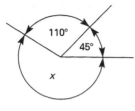

2 Find the angle x in Figure 9.17. The sum of the angles at a point is $360°$

$$360° - (110° + 45) = 360° - 155° = 205°$$

Answers to exercises

9.1 As shown in Figure 9.18, the starting point, $0°$, and the finishing point, $360°$, occupy the same position.

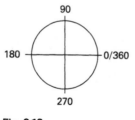

Fig. 9.18

Summary

We have been looking at:

- Measuring angles.
- Various types of angles and their properties.

Things to remember

- At the point where two lines intersect both internal and external angles are formed; this common point is called the vertex. The difference in direction of the two lines is referred to and measured as the angle between two lines. The length of the lines has no bearing on the angle.
- Angles can be measured and drawn using either a set square or a protractor. There are two types of commercially produced protractors: circular and semi-circular.

- The three points, the point at the vertex and a point on each of the two half lines, arc lettered and the angle referred to by these. Often a small angle / is put before the letters as shorthand for 'the angle'. Right angles or 90° angles are often marked with a small square in the angle.
- An acute angle is an angle that is less than 90°.
- An obtuse angle is one that is greater than 90° but less than 180°. (Remember, 180° would be a straight line.)
- Reflex angles are angles that are greater than 180° and less than 360°. (Remember, 360° would be a complete circle.)
- For any two angles that lie on a straight line, the sum of the two angles must be 180°.
- The sum of the angles at a point forms a complete revolution (360°).

Take a break for you studies before testing your skills on the following self-assessment questions.

Self assessment questions

Below there are three questions. Take your time in answering them.

- These are not meant as a test; the questions are simply to help you learn.
- Look back at you own notes and Chapter 9 if you need help.
- Answers and comments follow the questions, but you should look at them only when you have finished or are really stuck.

SAQ 9.1   Using a protractor, draw the following angles, indicating whether each angle is acute, obtuse or reflex:

12°, 120°, 300°, 103°, 45°, 246°, 27°, 170°, 230°.

SAQ 9.2   Find the supplementary angle x in Figure 9.19.

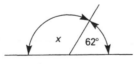

Fig. 9.19

SAQ 9.3   Find the angle x in Figure 9.20.

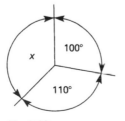

Fig. 9.20

Answers and comments

SAQ 9.1

Acute angles are less than 90°.

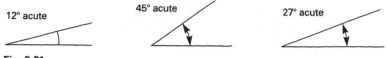

Fig. 9.21

Obtuse angles are greater than 90° but less than 180°.

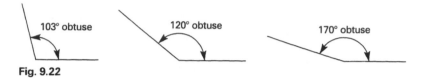

Fig. 9.22

Reflex angles are greater than 180° but less than 360°.

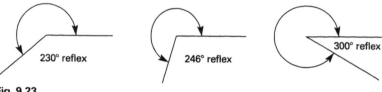

Fig. 9.23

SAQ 9.2

The sum of two angles that lie on a straight line is 180°. By subtracting the angle of 62° from 180°, x can be found.

$180° - 62° = 118°$
$x = 118°$

SAQ 9.3

The sum of the three angles is 360°. By adding the two angles 100° and 110° and subtracting the result from 360°, the angle x can be found.

$100° + 110° = 210°$
$360° - 210° = 150°$
$x = 150°$

CHAPTER 10 Triangles and quadrilaterals

> *Prior Knowledge* If you feel you are already skilled in the subjects mentioned above, turn to page 126 and try working through the SAQs. If you find difficulty in answering them, work through this chapter.

10.1 Plane figures

A **plane figure** is a figure that is flat – that is, it has no depth. Its shape is made from the lines that form the sides of the figure.

10.2 Triangles

The triangular shape is widely employed in the construction industry as it not only offers great strength and flexibility of design but is also relatively easy to construct. (Try to think where the triangular shape is used in your trade).

A triangle is an area, on a flat surface, enclosed by three straight lines that intersect at three points, shown as A, B and C in Figure 10.1. An angle is formed where any two lines meet. The amount of opening between the lines dictates the angles that make up the triangle. The angles of a triangle are not dependent on the lengths of the triangle's sides or the area it covers.

Compare the two triangles in Figures 10.1 and 10.2: notice that although the triangle in Figure 10.1 is much smaller than the triangle in Figure 10.2 (its sides are not as long and it does not cover such a large area) it has the same angles. Triangles that share the same **corresponding** angles (i.e. at the same positions in the triangles) are called **similar** triangles.

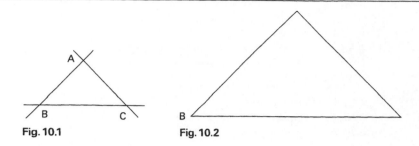

Fig. 10.1   **Fig. 10.2**

Triangles are called after the letters indicating the lines that form them – for example, the triangle shown in Figure 10.1 would be know as the **triangle ABC** as it is constructed from the lines AB, BC and CA. It is not important in which order the letters of the triangle are given; ABC or BCA or CAB or ACB or CBA or BAC all refer to the same triangle. A small triangle △ is often put before the letters as shorthand for 'the triangle'. The **perimeter** of a triangle can be found by adding together the lengths of the three sides that make up the triangle.

10.3   The angles in a triangle

The notation for angles we looked at in Chapter 9 can be used to name the angles in a triangle; for example, in Figure 10.3, ∠BAC in △ABC is the angle at point A. This is often referred to as angle A or ∠A.

The sum of the angles of a triangle is always 180°; therefore, if two of the angles are known the third can be found.

Example

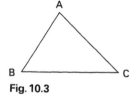

Fig. 10.3

1   If in the triangle shown in Figure 10.3 ∠A = 70° and ∠B = 60°, what is ∠C?

$$\angle C = 180° - (70° + 60°)$$
$$= 180° - 130°$$
$$= 50°$$

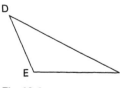

Fig. 10.4

2   In △DEF shown in Figure 10.4, ∠D = 42° and ∠F = 36°. What is ∠E?

$$\angle E = 180° - (42° + 36°)$$
$$= 180° - 78°$$
$$= 102°$$

Note: You can validate your answers by summing the angles. The result should be 180°.

Exercise 10.1

Find the third angle of a triangle if the other two angles are:

(a)   20° and 40°
(b)   75° and 30°

(c)   45° and 45°

(d)   50° and 50°

You should also get a protractor, and practise these triangles.

10.4 Types of triangles

Fig. 10.5

(a)   A **right-angled** triangle has one of its angles set at 90°. Figure 10.5 is a right-angled triangle, its longest side, called the **hypotenuse**, is always opposite the right angle.

(b)   In **acute-angled** or **scalene** triangle (Figure 10.6) every angle is less than 90°.

(c)   If one of the angles in a triangle is greater than 90° (as x is in Figure 10.7) then it is an **obtuse-angled** triangle.

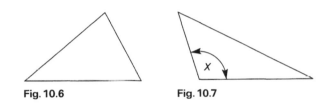

Fig. 10.6   Fig. 10.7

(d)   An **isosceles** triangle (Figure 10.8) has two sides, AB and AC, that are equal in length and two angles, x and y, that are equal. The two equal angles lie opposite the equal sides. A single line (sometimes a double line) is drawn through the two sides to show they are equal.

(e)   An **equilateral** triangle, as shown in Figure 10.9, has three sides that are equal in length (AB, BC and CA) and three angles of equal size (x, y and z. The angles are all 60° ($180 \div 3 = 60$). A single line (sometimes a double line) is drawn through the three sides to show they are equal. (*Note*: The word *equilateral* is Latin for 'equal sided'.)

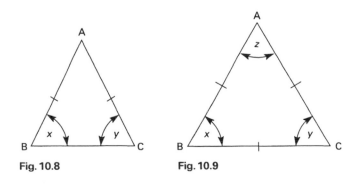

Fig. 10.8   Fig. 10.9

Exercise 10.2

What is the size of the angles marked x and y in Figure 10.10?

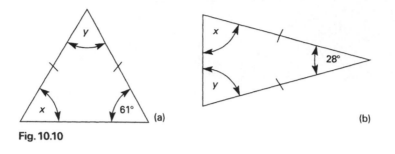

Fig. 10.10

10.5 ## Pythagoras' theorem

First a brief history lesson. Pythagoras was born on the Greek island of Samos at the beginning of the sixth century BC. At the age of 22 he belonged to and led a group of mathematicians who, among other things, worked out mathematical rules and theorems. The theorem we are interested in is his 45th, concerning right-angled triangles, in which he states that '*in a right-angled triangle, the square on the hypotenuse is equal to the sum of the squares on the two other sides.*'

Before we look closely at his theory, let's go further back in time and history. One of the civilisations before the Greeks was the Egyptians, who were ruled by a type of king called a Pharaoh. The Egyptians, who even today must be thought of as among the world's best builders, built the pyramids as burial tombs for their kings and queens, and their treasure.

The pyramids were built to a strict plan, a square base, with each corner being a true right angle and four sloping sides, each side an isosceles triangle. It has been estimated that some of the 82 pyramids built contained 2,000,000 blocks of stone, each having an average weight of 2.5 tonnes. Even if this task were done today, it would present problems to engineers, yet it was carried out with none of the sophisticated surveying and levelling equipment we now possess. Among the many problems they had to overcome was how to ensure that the base of the pyramids and the stones were square. They used a method to find a right angle that is still in use today, and that Pythagoras theorised much later, known in the building trade as 3, 4, 5.

Below I have explained the method the Egyptians used; you should try it for yourself before going on to the mathematical explanation.

Note: At the time of the Pharaohs the measurement in use, the **cubit**, was said to be the length of the Pharaoh's forearm. As there were a succession of Pharaohs there was probably some confusion about the length of the cubit.

For our purposes the units can be metres. The Egyptians would take a piece of rope and tie knots in it at 1 cubit intervals along its length. Two pegs would be hammered into the ground (A and B) four cubits apart. At roughly a right angle to AB and three cubits apart, a third peg (C) would be held. The third peg would be moved until the rope showed exactly five cubits between pegs A and C and three cubits between pegs B and C. The angle ABC, as shown in Figure 10.11 is a right angle. The larger the units are, the more accurate the angle.

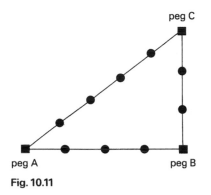

Fig. 10.11

Pythagoras used his 45th theorem to give mathematical terms to the 3, 4, 5 method of obtaining a true right angle; he gave it in an algebraic formula that said $a^2 + b^2 = c^2$, where a, b and c are the sides of a triangle. (Remember that the small raised 2 means that a number is to be squared, or multiplied by itself.)

Let us now give numerical values to a, b and c. If $a = 3$ and $b = 4$, the equation now becomes

$$3^2 + 4^2 = c^2$$

or

$$(3 \times 3) + (4 \times 4) = c^2$$

which in turn becomes

$$9 + 16 = c^2$$

We can now say that $c^2 = 25$. However, we really want to know the value of c. We can find this by finding the square root of 25 (look back to Chapter 6 if you have a problem with square roots).

I hope you found the answer 5 by either looking back to the table in Chapter 6 or by using your calculator.

The position of a and b can be switched, but c is always the longest side in the right angled triangle, the hypotenuse. We are often given the measurements for a and b and need to find c.

Example

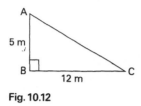

Fig. 10.12

The △ABC (Figure 10.12) has a right angle at B. The length of AB is 5 m and the length of BC is 12 m. Find the length of AC, the hypotenuse.

$$AC^2 = AB^2 + BC^2$$
$$= 5^2 + 12^2$$
$$= 25 + 144$$
$$= 169 \text{ m}^2$$

Therefore

$$AC = \sqrt{169} = 13$$

The length of AC is 13 m.

Notice that I am now referring to the sides of the triangle by the letters that indicate the lines that make up the triangle. So

$$a^2 + b^2 = c^2 \text{ has now become } AC^2 = AB^2 + BC^2$$

If the length of the hypotenuse (the longest side) in a right-angled triangle is given and the length of one other side is known, then by rearranging the equation $AC^2 = AB^2 + BC^2$ to $AB^2 = AC^2 - BC^2$ or $BC^2 = AC^2 - AB^2$ the length of the unknown side can be found.

Example

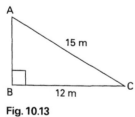

Fig. 10.13

The length of the hypotenuse, AC in the triangle ABC (Figure 10.13) is 15 m and the length of the side BC is 12 m. Find the length of the side AB.

$$AB^2 = AC^2 - BC^2$$
$$= 15^2 - 12^2$$
$$= 225 - 144$$
$$= 81 \text{ m}$$

Therefore

$$AB = \sqrt{81} = 9 \text{ m}$$

Pythagoras' theorem can also be used to find the height of isosceles and equilateral triangles, providing the lengths of the base and one side are known. (You will need the use of a calculator for the following examples and exercises.)

Example

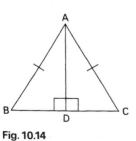

Fig. 10.14

What is the height of the triangle ABC if AB = 10 mm and BC = 8 mm?

The isosceles triangle ABC in Figure 10.14 can, by drawing a line AD from A to the middle point of BC, form two identical right-angled triangles, BDA and ADC. To find the height of the triangle we need to find the length of AD.

The line AB = 10 mm and the line BD = 4 mm (half of BC).

$$AD^2 = AB^2 - BD^2$$
$$= 10^2 - 4^2$$
$$= 100 - 16$$
$$= 84$$

Therefore

$$AD = \sqrt{34} = 9.165 \text{ mm (correct to three decimal places)}$$

 ### Exercise 10.3

Find the height of the following triangles, correct to three decimal places:

(a)   An isosceles triangle with two equal sides of 10 mm and a base of 12 mm.
(b)   An equilateral triangle with sides of 8 m.
(c)   An isosceles triangle with sides of 7 m and a base of 8 m.
(d)   An equilateral triangle with sides of 6 m.

10.6 Congruency

If two triangles are identical and will fit perfectly over one another, coinciding at all points, then they are said to be **congruent** triangles.

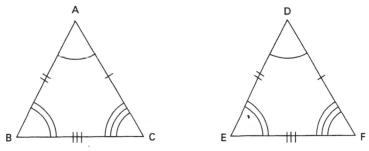

Fig. 10.15

The two triangles ABC and DEF in Figure 10.15 are congruent as

AB = DE	∠A = ∠D
BC = EF	∠B = ∠E
CA = FD	∠C = ∠F

They are identical to each other, coinciding at all points and have the same area. There are then seven elements – the three sides, the three angles and the area – that must be the same in both triangles before they can be called

congruent. However, not all seven elements need to be compared to *prove* congruency; if any of the following conditions for two triangles are met, then the triangles will be congruent.

(a)   The three sides of one triangle are equal in length to the three sides of the second triangle, as in Figure 10.16.

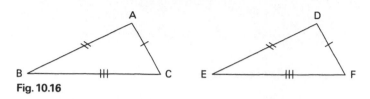

Fig. 10.16

(b)   Two of the sides and the angle between them in one triangle are equal in all respects to the corresponding sides and angle in the second triangle, as in Figure 10.17.

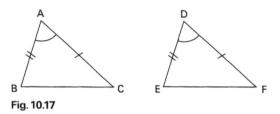

Fig. 10.17

(c)   One side and two angles in one triangle are equal in all respects to the corresponding side and angles in the second triangle, as in Figure 10.18.

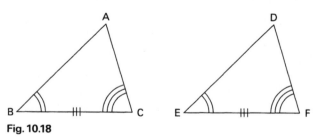

Fig. 10.18

(d)   The hypotenuse and one other side of a right-angled triangle are equal in all respects to the hypotenuse and corresponding side in the second triangle, as in Figure 10.19.

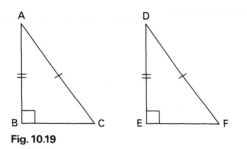

Fig. 10.19

10.7 Quadrilaterals

A quadrilateral is a figure bounded by four straight lines, as in Figure 10.20.

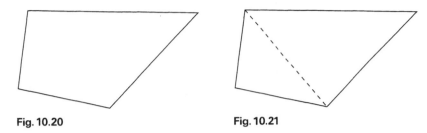

Fig. 10.20   **Fig. 10.21**

It can be split into two triangles with a diagonal line, as in Figure 10.21. The sum of the angles in a triangle is 180°, therefore the sum of the four angles in a quadrilateral must be 360°. If three of the angles are known, the fourth can be found.

Example

In the quadrilateral shown in Figure 10.22, find the size of the angle x.

We know the sum of the angles ABCD is 360°, therefore

$$x = 360° - (82° + 78° + 75°)$$
$$= 360° - 235°$$
$$= 125°$$

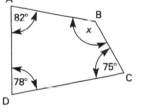

Fig. 10.22

The following are all examples of quadrilaterals.

Parallelogram

In a parallelogram the sides that are opposite each other are of equal length and are parallel to each other (Figure 10.23). The angles that are opposite to each other are also equal. Providing one of the angles is known, the other three can be found. The diagonals AD and BC bisect each other and the parallelogram to form two congruent triangles. The area of a parallelogram can be found by multiplying the base by the **perpendicular** height (this means at *right angles* to the base).

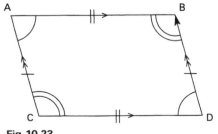

Fig. 10.23

Example

Find the area and the angles *x*, *y* and *z* for the parallelogram in Figure 10.24.

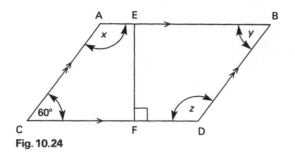

Fig. 10.24

Area: If the base CD = 6 m and the line EF = 3 m

$$Area = CD \times EF$$
$$= 6 \text{ m} \times 3 \text{ m}$$
$$= 18 \text{ m}^2$$

Angles: As the angle ACD equals 60°, then the angle opposite *y* must also equal 60°. The sum of the four angles is equal to 360° and angles *x* and *z* are opposite and equal. Therefore

$$x + z = 360° - (60° + 60°)$$
$$= 240°$$

and as they are equal

$$240° \div 2 = 120°$$
$$x = 120°$$
$$z = 120°$$

Rectangle

In a rectangle all the angles are right angles, as shown in Figure 10.25. The two sides opposite each other are parallel and equal in length. The two diagonals AD and CB are equal in length and bisect each other. The diagonals bisect the rectangle to form two congruent triangles. The method of finding the area of a rectangle is covered in Chapter 7.

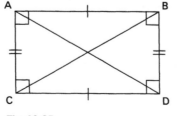

Fig. 10.25

Square

A **square** (Figure 10.26) is a rectangle with all its sides equal in length. The two diagonals AD and CB are equal in length and bisect each other. The diagonals intersect at 90° and bisect the square to form congruent triangles. The area of a square $= L^2$, where L is the length of one side of the square.

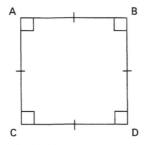

Fig. 10.26

Note: The fact that two diagonals of a rectangle or a square are equal in length can be used to good effect for testing for squareness a variety of building components, from the building lines of a new building to the door frames to be fitted in it (see Figure 10.27). If the two diagonals are of equal length then the corners of the building or door frame are at 90°

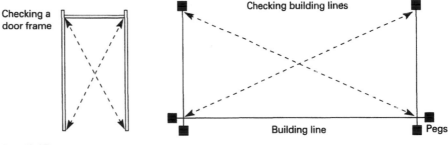

Fig. 10.27

Rhombus

A rhombus has all the properties of a parallelogram except that its sides are of equal length. The diagonals AD and BC bisect at right angles and divide the rhombus into four congruent triangles. The area of a rhombus can be found by multiplying the base by the perpendicular height, as we did for the parallelogram.

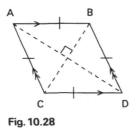

Fig. 10.28

Trapezium

A trapezium is a quadrilateral that has two unequal parallel sides. It can be divided with a diagonal into two triangles. The area of the trapezium can be calculated as the sum of the areas of these two triangles, providing the height and the lengths of the two parallel sides are known.

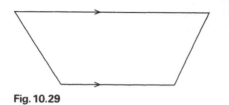

Fig. 10.29

Example

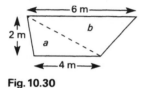

Fig. 10.30

Find the area of the trapezium shown in Figure 10.30. The trapezium has been divided into two triangles, *a* and *b*.

$$\text{Area of triangle } a = \frac{2 \times 4}{2} = 4 \text{ m}^2$$

$$\text{Area of triangle } b = \frac{2 \times 6}{2} = 6 \text{ m}^2$$

Area of trapezium = Area of triangle *a* + Area of triangle *b*

$$= 4 \text{ m}^2 + 6 \text{ m}^2$$

$$= 12 \text{ m}^2$$

(Look back to Chapter 7 now if you have any doubts about the method used for finding the area of a triangle.)

 ### Exercise 10.4

(a) Find the angles *x* and *y* in the quadrilaterals shown in Figure 10.31.

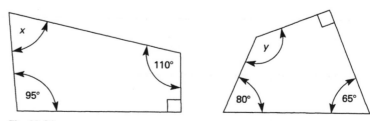

Fig. 10.31

(b) Find the angles x, y and z and the area of the parallelogram shown in Figure 10.32.

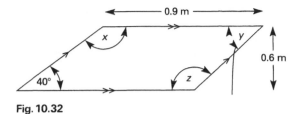

Fig. 10.32

(c) Find the area of the building plot shown in Figure 10.33.

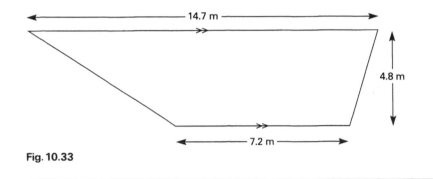

Fig. 10.33

|10.8| Polygons

Any shape whose sides are constructed with straight lines is called a **polygon** (the word *polygon* is Greek for 'many sided'). Both triangles and quadrilaterals are polygons; the triangle is a three-sided polygon and the quadrilateral is a four-sided polygon.

If the sides and angles of a polygon are all the same size, it is called a **regular** polygon. The square and the equilateral triangle are three- and four-sided regular polygons.

All regular polygons can be divided into a number of congruent isosceles triangles. If the size and the number of the triangles are known, then the area of the polygon can be found.

A regular **pentagon**, Figure 10.34, is a polygon that has five sides and contains five congruent isosceles triangles, as the triangle ABC.

The angle ABC can be found

$$\angle ABC = 360° \div 5$$
$$= 72°$$

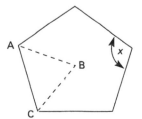

Fig. 10.34

The angles CAB and ACB can be found

$$\angle CAB = 180° - 72°$$
$$= 108° \div 2 = 54°$$
$$\angle ACB = 180° - 72°$$
$$= 108° \div 2 = 54°$$

The angle marked x is called an interior angle and within a regular polygon it can be found by using the formula

$$\frac{(2n - 4) \times 90}{n}$$

where n is equal to the number of sides on the regular polygon.

Example

Find the angle x in Figure 10.34.

As Figure 10.34 is a pentagon and has five sides, $n = 5$. Substituting 5 for n in the formula $\dfrac{(2n - 4) \times 90}{n}$ we get

$$x = \frac{((2 \times 5)) - 4) \times 90}{5}$$
$$= \frac{(10 - 4) \times 90}{5} = \frac{6 \times 90}{5} = \frac{540}{5} = 108°$$

Other regular polygons are:

- A hexagon, which has 6 sides.
- A heptagon, which has 7 sides.
- An octagon, which has 8 sides.
- A nonagon, which has 9 sides.
- A decagon, which has 10 sides.
- An undecagon, which has 11 sides.
- A duodecagon, which has 12 sides.

Exercise 10.5

Use the formula $\dfrac{(2n - 4) \times 90}{n}$ to find the internal angles of:

(a) A three-sided regular polygon.
(b) A four-sided regular polygon.
(c) A six-sided regular polygon.
(d) A seven-sided regular polygon.

Answers to exercises

10.1   (a) 120°; (b) 75° (c) 90° (d) 80°

10.2 (a) $x = 61°$, $y = 58°$; (b) $x = 76°$, $y = 76°$

10.3 (a) 8 mm; (b) 6.928 m; (c) 5.745 m; (d) 5.196 m

10.4 (a) $x = 65°$, $y = 125°$; (b) $x = 140°$, $y = 40°$, $z = 140°$, area $= 54$ mm^2
(c) 52.56 m^2

10.5 (a) 60°; (b) 90°; (c) 120°; (d) 128.57°

Summary

We have been looking at:

- Triangles.
- The angles of a triangle.
- Types of triangles.
- Pythagoras' theorem.
- Congruency.
- Quadrilaterals.
- Polygons.

Things to remember

- A triangle is an area, on a flat surface, enclosed by three straight lines that intersect at three points. Where two lines meet, an angle is formed. The amount of opening between the lines dictates the angles that make up the triangle. Triangles are called after the letters indicating the lines that form the triangle.
- It is not important in which order the letters of the triangle are given. A small triangle is often put before the letters as shorthand for 'the triangle'.
- The sum of the angles in a triangle is always 180°.
- A right-angled triangle has one of its angles set at 90°. Its longest side, called the hypotenuse, is always opposite the right angle.
- An acute-angled or scalene triangle has every angle less than 90°.
- If one of the angles in a triangle is greater than 90°, then it is an obtuse-angled triangle.
- An isosceles triangle has two sides that are equal in length and two angles that are equal. The two angles that are equal lie opposite the equal sides. A single line (sometimes a double line) is drawn through the sides to show they are equal.
- An equilateral triangle has three sides that are equal in length and three angles of equal size. The angles are 60° (180° ÷ 3 = 60°).
- Pythagoras' theorem states that '*the square on the hypotenuse of a right-angled triangle is equal to the sum of the squares on the two other sides.*' The theorem confirms that a triangle with sides 3, 4 and 5 units in

length is a right-angled triangle. It can also be used to find the height of isosceles and equilateral triangles, providing the lengths of the base and one side are known.

- If two triangles are identical to each other and will fit perfectly one over another, coinciding at all points, then they are said to be congruent. Not all seven elements of a triangle need to be compared to prove congruency.
- A quadrilateral is a figure bounded by four straight lines. It can be split into two triangles with a diagonal line. The sum of the four angles in a quadrilateral is 360°. If three of the angles are known the fourth can be found.
- Any shape whose sides are constructed with straight lines is called a polygon. If the sides and angles of a polygon are all the same size it is called a regular polygon.
- The interior angle of a regular polygon can be found using the formula

$$\frac{(2n - 4) \times 90}{n}$$

where n is equal to the number of sides on the regular polygon.

Take a break from you studies before testing your skills on the following self-assessment questions.

Self-assessment questions

Below there are eight questions. Take your time in answering them.

- These are not meant as a test; the questions are simply to help you learn.
- Look back at you own notes and Chapter 10 if you need help.
- Answers and comments follow the questions, but you should look at them only when you have finished or are really stuck.

SAQ 10.1   Fill in the missing angle for the triangles listed below

1st angle	2nd angle	3rd angle
12°	36°	
38°	38°	
42°	45°	
27°	27°	

SAQ 10.2   Find the size of the angles x and y in the triangles (a) and (b) in Figure 10.35.

SAQ 10.3   What size are the angles x, y and z in $\triangle$ ABC shown in Figure 10.36?

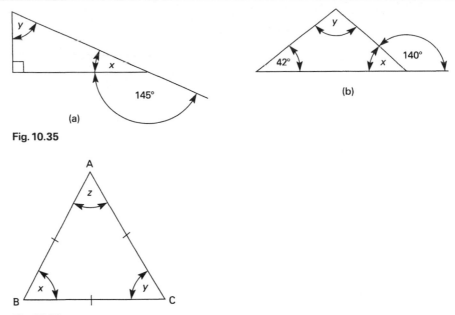

(a)

(b)

Fig. 10.35

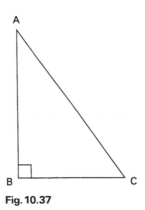

Fig. 10.36

SAQ 10.4   The right-angled triangle ABC shown in Figure 10.37 has a side AB = 8 m and a side CB = 6 m.

(a)   What is the total length of the perimeter of the triangle?

(b)   What is the area of the triangle?

(Look back to Chapter 7 if you cannot remember how to find the area of a triangle.)

Fig. 10.37

SAQ 10.5   Find the height of an isosceles triangle with a base of 14 m and two equal sides 11 m in length. Give your answer correct to three decimal places.

SAQ 10.6   Find the height of an equilateral triangle that has sides 4 m in length. Give your answer correct to two decimal places.

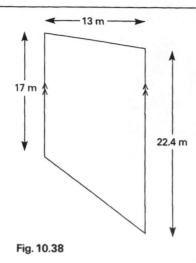

Fig. 10.38

SAQ 10.7 What is the area of the building plot shown in plan view in Figure 10.38?

SAQ 10.8 Complete the table:

The interior angle of a regular pentagon is 108°
The interior angle of a regular hexagon is
The interior angle of a regular heptagon is
The interior angle of a regular octagon is
The interior angle of a regular nonagon is
The interior angle of a regular decagon is
The interior angle of a regular undecagon is
The interior angle of a regular duodecagon is

Answers and comments

SAQ 10.1

The sum of the angles in a triangle is always 180°, and so by summing the 1st and 2nd angles and subtracting the result from 180° the 3rd angle can be found:

1st angle	2nd angle	3rd angle
12°	36°	132°
38°	38°	104°
42°	45°	93°
27°	27°	126°

SAQ 10.2

(a) The small square tells us that (a) is a right-angled triangle and so must be 90°. As it lies on a straight line with the angle 145° the size of the angle x can be found by

$$x = 180° - 145°$$
$$= 35°$$

The sum of the angles of a triangle must add up to 180°, and we now know that two of the angles are 90° and 35°.

$$y = 180° - (90° + 35°)$$
$$= 180° - 125°$$
$$= 55°$$

Check the answer: $90° + 35° + 55° = 180°$

55°

(b) The angle x can be found as it lies on a straight line with the angle 140°

$$x = 180° - 140°$$
$$= 40°$$

The sum of the angles of a triangle must add up to 180°, and we know that two of the angles are 40° and 42°.

$$y = 180° - (40° + 42°)$$
$$= 180° - 82°$$
$$= 98°$$

Check the answer: $40° + 42° + 98° = 180°$

SAQ 10.3

The triangle ABC in Figure 10.39 is an equilateral triangle; the three angles x, y and z are of equal size 60° (180 ÷ 3 = 60). The lines drawn through the three sides shows they are of equal length.

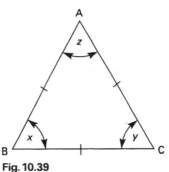

Fig. 10.39

SAQ 10.4

(a) The perimeter of the triangle in Figure 10.40 can be found by adding together the lengths of the three sides. The lengths of two of the sides are given as AB = 8 m and CB = 6 m. Before we can find the

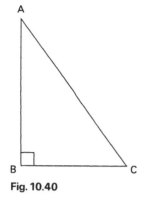

Fig. 10.40

perimeter of the triangle the length of AC has to be found, using Pythagoras' theorem.

$$AC^2 = AB^2 + CB^2$$
$$= 8^2 + 6^2$$
$$= 64 + 36$$
$$= 100$$
$$= \sqrt{100}$$
$$= 10 \text{ m}$$

The perimeter can now be found by adding the three sides of the triangle.

$$10 \text{ m} + 8 \text{ m} + 6 \text{ m} = 24 \text{ m}$$

(b)   The area of a triangle can be found by multiplying its base by its perpendicular height and dividing the result by 2:

$$\begin{array}{ll} 6 \text{ m} & \text{Base (CB)} \\ \times \ \underline{8 \text{ m}} & \text{Perpendicular (AB)} \\ 48 \text{ m}^2 & \end{array}$$

Area of triangle $= 48 \text{ m}^2 \div 2 = 24 \text{ m}^2$

SAQ 10.5

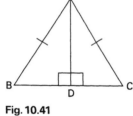

Fig. 10.41

I have drawn the isosceles triangle ABC as Figure 10.41 to help in the explanation of how the answer is arrived at. To find the height of the triangle first find the length of AD:

$$AD^2 = AB^2 - BD^2$$
$$= 11^2 - 7^2 \quad \text{(BD is half the base BC of 14 m)}$$
$$= 121 - 49$$
$$= 72$$

$$AD = \sqrt{72}$$
$$= 8.485 \quad \text{(correct to three decimal places)}$$

Height of triangle $= 8.485$ m.

SAQ 10.6

Again we need to find the length of a line AD drawn from point A to the middle point of BC. (I have not drawn the triangle this time as I am sure you are familiar with the lines and points.)

$$AD^2 = AB^2 - BD^2$$
$$= 4^2 - 2^2 \quad \text{(half the base 4 m)}$$
$$= 16 - 4$$
$$= 12$$

$$AD = \sqrt{12}$$
$$= 3.46 \text{ m} \quad \text{(correct to two decimal places)}$$

Height of triangle $= 3.46$ m.

SAQ 10.7

The area of the building plot in Figure 10.42 can be found by dividing it into two triangles (a) and (b), as shown in Figure 10.40, finding the area of the two triangles, and adding the two areas together.

Area of triangle (a)

$$\frac{13 \text{ m} \times 17 \text{ m}}{2} = 110.5 \text{ m}^2$$

Area of triangle (b)

$$\frac{13 \text{ m} \times 22.4 \text{ m}}{2} = 145.6 \text{ m}^2$$

Area of building plot is

$$110.5 \text{ m}^2 + 145.6 \text{ m}^2 = 256.1 \text{ m}^2$$

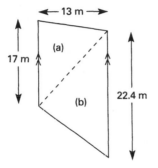

Fig. 10.42

SAQ 10.8

Using the formula $\dfrac{(2n - 4) \times 90}{n}$, where n is the number of sides in the polygon, gives the following results. (*Note*: I have put the number of sides in the polygon after the size of the interior angle.)

The interior angle of a regular pentagon is	108°	(5)
The interior angle of a regular hexagon is	120°	(6)
The interior angle of a regular heptagon is	128.57°	(7)
The interior angle of a regular octagon is	135°	(8)
The interior angle of a regular nonagon is	140°	(9)
The interior angle of a regular decagon is	144°	(10)
The interior angle of a regular undecagon is	147.27°	(11)
The interior angle of a regular duodecagon is	150°	(12)

11.1 Evaluating formulae
11.2 Transposing formulae

Prior Knowledge If you feel you are already skilled in the subjects mentioned above, turn to page 136 and try working through the SAQs. If you find difficulty in answering them, work through this chapter.

11.1 Evaluating formulae

A **formula** is a method of expressing the value of a quantity in terms of its relationship with other values directly associated with it. We have already looked at and used several formulae; for example, in Chapter 7 the formula $A = L \times W$ was employed to evaluate the area of a rectangle. The expression $A = L \times W$ is a formula for A in terms of L and W.

The value of A (the area) can be evaluated after substituting the given values for L (the length) and W (the width) and some simple mathematics.

Examples

1 Use the formula $A = L \times W$ to find the value of A if $L = 4$ and $W = 5$.

Substitute the values given for L and W:

$A = 4 \times 5 = 20$

2 Use the formula $A = L \times W$ to find the value of A if $L = 27.3$ m and $W = 14.345$ m.

Substitute the values given for L and W:

$A = 27.3 \text{ m} \times 14.345 \text{ m} = 391.6185 \text{ m}^2$

Formulae can be thought of as a kind of maths machine; you put in the values and out pops the answer, as you would with a calculator: [4] [×] [5] [=] 20. The formulae used by students of the building trade fall broadly into two categories:

(a) The kind that could be used every day at work, that you would be expected to remember and use as easily as you would tools from your own tool kit. Those listed below were first encountered in Chapter 7.

$$A = \text{Length} \times \text{Width} \qquad \text{area of a rectangle}$$

$$A = \text{Base} \times \text{Perpendicular height} \div 2 \qquad \text{area of a triangle}$$

$$C = 2\pi r \qquad \text{circumference of a circle}$$

$$A = \pi r^2 \qquad \text{area of a circle}$$

$$V = L \times W \times D \qquad \text{volume or capacity of cuboid}$$

$$V = \pi r^2 h \qquad \text{volume or capacity of a cylinder}$$

(b)   The type you will not be expected to remember, but should know where to find and how to use to solve problems. Many of these at first glance can seem quite daunting; however, all that is usually required is for you to substitute the values you have for the letters or symbols in the formula and, with the use of a calculator, carry out a relatively simple mathematical task.

Example

The formula below is used for finding the discharge in litres per second for drainage pipes:

$$q = \sqrt{\frac{H \times d^5}{25 \times L \times 10^5}}$$

where q = discharge in litres per second
   H = head in metres
   L = length of pipe in metres
   d = internal diameter of drain pipe in millimetres

How many litres a second would 30 metres of 50 mm pipe with an effective head of 4 metres discharge?

Substituting the letters in the formula for the quantities given

$$q = \sqrt{\frac{4 \times 50^5}{25 \times 30 \times 10^5}}$$

Dividing the bottom figure into the top figure

In scientific notation this becomes

$$q = \sqrt{\frac{1.25 \times 10^9}{0.75 \times 10^8}} = \sqrt{16.666667} = 4.0824829$$

say 4.08 litres per second

11.2 Transposing formulae

Often in calculations, in order that a different value can be found, the formula given is required to be rearranged before evaluation can take place. For example, the formula $A = L \times W$ (used for finding the area of a rectangle) has *A* as the subject – that is, by multiplying together *L* and *W*, *A* can be found. Suppose, however, we already know what figures *A* and *W* represent and are required to find *L*. This method of rearranging the formula so that one of the other letters or symbols becomes the subject is called **transposing**.

It is often useful when transposing a formula to think of it as being balanced about the equals sign, as in Figure 11.1, and that whatever is done to terms on one side of the equals sign must also be done to the terms on the other side to maintain that balance.

A letter or symbol that is moved from one side of the equals sign to the other side changes its sign, for example

- Division changes to multiplication.
- Multiplication changes to division.
- Addition changes to subtraction.
- Subtraction changes to addition.
- Roots change to powers.
- Powers change to roots.

Fig. 11.1

Examples

1 Transpose the formula $A = L \times W$ to make *L* the subject.

$$A = L \times W$$ *W* needs to be moved to the other side of the equal sign leaving *L* to become the subject.

$$\frac{A}{W} = \frac{L \times W}{W}$$ Divide both sides by *W* to maintain the balance, as in Figure 11.2.

The two *W*s cancel each other out to leave

$$\frac{A}{W} = L$$ still in balance.

Fig. 11.2

This rearrangement can be used to find the length if we know the area and the width of a rectangle.

2 Transpose the formula for finding the area of a triangle ($A = W \times H \div 2$) to find the height of a triangle that has an area of 12 m² and a base of 4 metres.

We first need to make *H* the subject of the formula $A = W \times H \div 2$. We move *W* to the other side of the equal sign by dividing both sides by *W* (notice how it is still in balance).

$$\frac{A}{W} = \frac{W \times H}{W} \div 2$$

Cancel out the *W*s to give

$$\frac{A}{W} = H \div 2$$

Multiplying both sides by 2 will remove the $\div 2$ from *H* to give

$$2\frac{A}{W} = H \quad \text{or alternatively} \quad H = 2(A \div W)$$

as either formula will give the same result.

We can now assign the values for A and W to find H:

$H = 2(12 \div 4)$
$\quad = 2 \times 3 = 6$

The height of the triangle is 6 metres.

3 What would be the height of a cylinder with a volume of 1.571 m^3 and a base with a diameter of 1 metre?

First make h the subject of the formula $V = \pi r^2 h$

$V = \pi r^2 h$ $\qquad$ divide each side by π to give

$\dfrac{V}{\pi} = \dfrac{\pi r^2 h}{\pi}$ $\qquad$ cancel out the π to give

$\dfrac{V}{\pi} = r^2 h$ $\qquad$ divide each side by r^2 to give

$\dfrac{V}{\pi r^2} = \dfrac{r^2 h}{r^2}$ $\qquad$ cancel out the r^2 to give

$\dfrac{V}{\pi r^2} = h$

The values for V and r can now be assigned. Remember that the radius r is half the diameter.

$$\frac{1.571}{0.5^2 \times \pi} = \frac{1.571}{0.5^2 \times 3.142} = \frac{1.571}{0.7855} = 2$$

The height of the cylinder is 2 metres.

4 Find the radius of a circle that has an area of 15 metres. Give the answer to three decimal places.

First make r the subject of the formula $A = \pi r^2$

$A = \pi r^2$ $\qquad$ divide each side by π to give

$\dfrac{A}{\pi} = \dfrac{\pi r^2}{\pi}$ $\qquad$ cancel out the π to give

$\dfrac{A}{\pi} = r^2$ $\qquad$ remembering that powers change to roots, gives

$\sqrt{\dfrac{A}{\pi}} = r$

The values for A and π can now be assigned.

$$\sqrt{\frac{15}{3.142}} = \sqrt{4.7756} = 2.1853146 = 2.185 \text{ m}$$

The radius of the circle is 2.185 metres.

There are no exercises for Chapter 11. Instead you should look through textbooks associated with your trade and practise evaluating and transposing the formulae found.

Summary

We have been looking at:

- Evaluating formulae.
- Transposing formulae.

Things to remember

- To evaluate a formula, substitute or replace the letters or symbols that make up the formula with the known values. Carry out the final calculation.
- When transposing a formula, a term that is transferred from one side of the equals sign to the other has its sign changed:
  - Division changes to multiplication
  - Multiplication changes to division
  - Addition changes to subtraction
  - Subtraction changes to addition
  - Roots change to powers
  - Powers change to roots
- When a term containing a root is squared the root is removed.

Take a break from your studies before testing your skills on the following self-assessment questions.

Self-assessment questions

Below there are eight questions. Take your time in answering them.

- These are not meant as a test; the questions are simply to help you learn.
- Look back at you own notes and Chapter 11 if you need help.
- Answers and comments follow the questions, but you should look at them only when you have finished or are really stuck.

SAQ 11.1 Using the following formula, find the value of w, where $f = 6$, $g = 9$ and $h = 18$:

$$w = \frac{f \times g^2}{h}$$

SAQ 11.2 The formula $4\pi r^2$ is used to find the area of a sphere. What would be the area of a sphere that had a diameter of 500 mm?

SAQ 11.3 Transpose the formula $f = \sqrt{3gd}$ to make d the subject.

SAQ 11.4 Transpose the formula $w = \dfrac{f \times g^2}{h}$ to make g the subject.

SAQ 11.5 If a rectangular room has a floor area of 18.9 m^2, what would the width be if the room was 4.2 metres in length?

SAQ 11.6 What length would the base of a triangle be that had an area of 14.5 m^2 and a height of 4 m?

SAQ 11.7 A cold water storage cistern has the capacity to hold 2,700 litres of water. It is 1.5 m wide and 900 mm from the bottom to the water line. What is its length?

SAQ 11.8 A circular swimming pool with a level base contains 78,550 litres of water when filled to the 1 metre level. What is the diameter of the pool?

Answers and comments

SAQ 11.1

Substitute the letters in the formula for the values given as $f = 6$, $g = 9$ and $h = 18$:

$$w = \frac{f \times g^2}{h} = \frac{6 \times 9^2}{18} = \frac{6 \times 81}{18} = \frac{486}{18} = 27$$

SAQ 11.2

Substitute the letters in the formula $4\pi r^2$ for the values given, remembering that the radius is half the diameter. I have expressed the millimetres as metres, i.e. 250 mm = 0.25 m.

Area of sphere = $4\pi r^2 = 4 \times 3.142 \times 0.25 \times 0.25 = 0.7855$ m^2.

SAQ 11.3

$f = \sqrt{3gd}$ square both sides to give

$f^2 = (\sqrt{3gd})^2$ the square root and the square cancel out to give

$f^2 = 3gd$ divide both sides by $3g$ to give

$\dfrac{f^2}{3g} = \dfrac{3gd}{3g}$ cancel out to give

$\dfrac{f^2}{3g} = d$

SAQ 11.4

$$w = \frac{f \times g^2}{h}$$ multiply both sides by h to give

$$hw = \frac{f \times g^2}{h} \times h$$ cancel out to give

$$hw = f \times g^2$$ divide both sides by f to give

$$\frac{HW}{f} = \frac{f \times g^2}{f}$$ cancel out to give

$$\frac{hw}{f} = g^2$$ square root both sides to give

$$\sqrt{\frac{hw}{f}} = \sqrt{g^2}$$ the square root and the square cancel out to give

$$\sqrt{\frac{hw}{f}} = g$$

SAQ 11.5

Transpose the formula $A = L \times W$ to make W the subject:

$A = L \times W$ divide both sides by L to give

$$\frac{A}{L} = \frac{L \times W}{L}$$ cancel out to give

$$\frac{A}{L} = W$$

Substitute the letters in the formula for the values given:

$$W = \frac{18.9}{4.2} = 4.5 \text{ metres}$$

SAQ 11.6

Transpose the formula $A = W \times H \div 2$ to make W the subject:

$A = W \times H \div 2$ divide both sides by H to give

$$\frac{A}{H} = \frac{W \times H}{H} \div 2$$ cancel out to give

$$\frac{A}{H} = W \div 2$$ multiply both sides by 2 to give

$$2\frac{A}{H} = 2(W \div 2)$$ cancel out to give

$$2\frac{A}{H} = W$$

Substitute the letters in the formula for the values given:

$$W = 2(14.5 \div 4) = 2 \times 3.625 = 7.25 \text{ metres}$$

SAQ 11.7

Transpose the formula $V = L \times W \times D$ to make L the subject:

$V = L \times W \times D$ divide both sides by $b \times D$ to give

$$\frac{V}{W \times D} = \frac{L \times W \times D}{W \times D} \quad \text{cancel out to give}$$

$$\frac{V}{W \times D} = L$$

Substitute the letters in the formula for the values given. I am expressing the 2,700 litres as a volume of 2.7 m^3.

$$L = \frac{V}{1.5 \times 0.9} = \frac{2.7}{1.35} = 2 \text{ metres}$$

SAQ 11.8

Transpose the formula $V = \pi r^2 h$ to make r the subject:

$V = \pi r^2 h$ divide both sides by πh to give

$\dfrac{V}{\pi h} = r^2$ cancel out to give

$\sqrt{\dfrac{V}{\pi h}} = r^2$ square root both sides to give

$\sqrt{\dfrac{V}{\pi h}} = \sqrt{r^2}$ the square root and the square cancel out to give

$\sqrt{\dfrac{V}{\pi h}} = r$

Substitute the letters in the formula for the values given. I am expressing the 78,550 litres as a volume of 78.550 m^3.

$$r = \sqrt{\frac{78.550}{3.142 \times 1}} = \sqrt{\frac{78.550}{3.142}} = \sqrt{25} = 5$$

Remember, this is the radius; to find the diameter the radius must be multiplied by 2.

$$\text{Diameter} = 5 \times 2 = 10 \text{ metres}$$

CHAPTER 12

Problem solving

Frequently when asked to solve a mathematical problem of the type found in this book or on a building site, difficulties can arise, not because of the often relatively simple calculations involved (especially if a calculator is used) but because you have not quite grasped what you are required to do. To this end it is vital that you

(a) Very carefully read and fully understand the problem, ensuring you have all the necessary information and the necessary skills required to solve all parts of the question, before starting.
(b) Set out your work in a clear, precise, neat, logical manner in well-defined easy-to-follow stages, labelling the parts of the solution as you go. If you use a calculator show how the answer you obtained was arrived at. This can be useful when you come to review your work some time later.
(c) Check your answer. Often in a work situation large quantities of materials, and hence large sums of money, can rest on the answer you give.

This chapter comprises ten mathematical problems of the type you may encounter while working in the building trade. Do not worry if they appear unconnected with your chosen trade; they are all based on what you have learnt in this book and could be thought of as a final assessment to see how far you have advanced with your mathematical skills.

The first two have been answered and are laid out as a guide for you. The notes in *italic* would not normally be included in the answer; they are there to help you.

Look back at you own notes and the relevant chapter if you find yourself stuck at any point. The answers to the remainder of the problems can be found on pages 145–8, where again some notes have been added in *italic*.

Often, as with many mathematical solutions, there are several ways in which an answer can be arrived at. Do not become over concerned if your method of finding the solution differs from mine, as long as the answers and the general method of laying out the answers agree.

Problems

Problem 1

A firm of wall tilers has been asked to submit a price for supplying and hanging tiles in twenty-six kitchens on a new housing estate. The tiles to be used cost £8 a square metre. The firm charges £6 a square metre of wall area to hang the tiles. An allowance of 10% is made on the total number of tiles required for cutting and wastage. Combined grout and adhesive costs £12 for a 15 kg tub which will cover approximately 16 square metres of wall area. The wall areas to be tiled are 3 m × 2 m, 2.2 m × 2.4 m and 1.7 m × 1.6 m.

Answer

Total area to be tiled of one kitchen:

1st wall	3 m × 2 m	=	6 m²
2nd wall	2.2 m × 2.4 m =		5.28 m²
3rd wall	1.7 m × 1.6 m =		+ 2.72 m²
			14 m²

Add these together to obtain tiled area of one kitchen.

Total area to be tiled of 26 kitchens:

14 m² × 26 = 364 m²   *To find the area for the twenty-six kitchens multiply the area of one by 26.*

Charge for hanging tiles:

364 m² at £6 a square metre
364 × £6 = £2,184

This is calculated before the 10% for cutting and wastage is added as the price is based on wall area.

Total and cost of tiles required:

364 m² + 10%
= 364 m² × 0.1 = 36.4 m²
= 364 m² + 36.4 m² = 400.4 m²
400.4 m² at £8 per square metre
= 400.4 × £8 = £3,203.20

Look back to Chapter 8 if you are not sure how to obtain percentages.

Adhesive and grout:

364 m² ÷ 16 = 22.75 tubs   *Again this is calculated on wall area. One tub of adhesive will hang and grout 16 m² of tiling.*

The tiler will have to price 23 tubs of adhesive:

23 tubs at £12 a tub
= 23 × £12 = £276.00

Totals

Cost of hanging tiles	£2,184.00
Cost of tiles	£3,203.20
Grout and adhesive	£276.00
Price to be submitted	£5,663.20

Add the totals together to obtain the final price.

A check should now be made on the answer. First look at the answer. Is it plausible? Does it make sense? Will anybody reading your solution be able to understand how you have laid it out and arrived at the answer? Check that you have done all that the problem asked of you and finally go back over and recheck your calculations.

Problem 2

Six concrete cylindrical columns, each with a diameter of 2 metres and a height of 5 metres, are to be demolished, broken up and carted away as rubble. The contractor has given a price of £37.50 a tonne for the job. If the concrete has a mass of 2,240 kg/m^3 how much will the contractor charge?

Answer

Volume of the columns using the formula $\pi r^2 h$

3.142×1 m $\times 1$ m $\times 5$ m $\times 6 = 94.26$ m^3 *By multiplying by 6 the total volume of all the columns can be found.*

One cubic metre of concrete has a mass of

$2,240 \div 1,000 = 2.240$ tonnes   *As the question is dealing with tonnes the mass of the concrete will be better expressed in tonnes per cubic metre.*

Total mass of concrete

94.26 m$^3 \times 2.240$ tonnes $= 211.1424$ tonnes

Cost

211.1424 tonnes at £37.50 a tonne
$= 211.1424 \times £37.50 = £7,917.84$

Again a check should be made on the answer. You could carry out a rough estimate on your answer, as we did in Chapter 3, to see if it is plausible.

£38 $\times 2$ tonnes $\times 94$ m$^3 = £7,144$

Problem 3

A sub-contract plumber is to tender for the installation of a central heating and domestic hot water system in six pairs of semi-detached houses. From the drawings the plumber has estimated that each of the houses requires 15 mm, 22 mm and 28 mm copper pipe to the following lengths.

15 mm	45 metres
22 mm	24 metres
28 mm	6 metres

Copper pipe is available in 3 metre lengths and the costs are as follows:

15 mm costs £3.50 per 3 metre length
22 mm costs £5.00 per 3 metre length
28 mm costs £7.50 per 3 metre length

What is the total cost of the copper pipe for the twelve houses?

Problem 4

The rafter shown in Figure 12.1 is for a roof pitched at 38°. The roof has a span of 5.2 metres and a rise of 2 metres. What, to the nearest millimetre, will be the length of the rafter and at what angles should it be cut at the ridge (*x*) and eaves (*y*)?

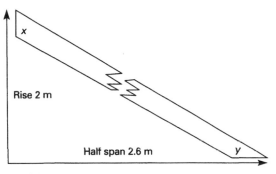

Rise 2 m

Half span 2.6 m

Fig. 12.1

Problem 5

A bricklayer has been asked to tender for the repointing to the rear of a chalet bungalow, as shown in Figure 12.2. The price is to include all materials and raking out the joints to a depth of 20 mm. The small window

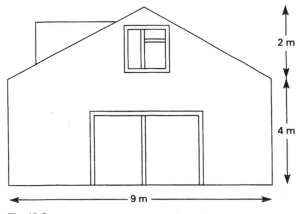

2 m

4 m

9 m

Fig. 12.2

in the gable measures 900 mm × 1.1 m and the patio window is 3 m wide and 2.1 m high.

The bricklayer charges £10 a square metre including materials for pointing and £2 a square metre for raking out. What will be the final estimate?

Problem 6

Twenty-four concrete window cills are to be cast in situ. All the cills have the same profile, as in Figure 12.3, but differ in length. The carpenters have finished the formwork and have written a list of the various lengths and have asked you, as the supervisor, to order the ready mixed concrete. Allowing 6% for spillage, how much to the nearest half a cubic metre should you order?

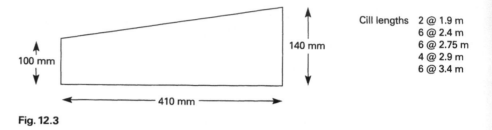

Cill lengths 2 @ 1.9 m
6 @ 2.4 m
6 @ 2.75 m
4 @ 2.9 m
6 @ 3.4 m

Fig. 12.3

Problem 7

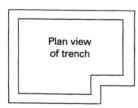

Plan view of trench

Fig. 12.4

A series of trenches 40 metres in total length have been excavated to a depth of 1 metre and a width of 750 mm as the footing of a new house (Figure 12.4). When removed the earth increased in bulk by 30%. The concrete and brickwork that make up the foundations will take up 48% of the trenches. How much of the earth should be removed from the site and how much should be retained for back filling?

Problem 8

It has been decided that the digging of the trenches in the previous problem is to be sub-contracted out. A gang of seven labourers have asked for £9 per man per hour to carry out the excavations and you have agreed to this.

Four labourers begin to dig the trenches at 8 a.m., and after three hours the other three labourers, who have been completing another job, join them. The whole gang takes a dinner break from 12 noon to 1 p.m., and all the labourers finish digging for the day at 6 p.m. What was the labour cost for digging that day?

Problem 9

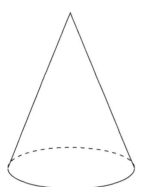

The volume of the cone shown in Figure 12.5 can be found by using the formula $V = \frac{1}{3}r^2h$. Transpose the formula to make the radius r the subject and find the diameter of a cone that has a volume of 4.713 m³ and a height of 2 m.

Fig. 12.5

Problem 10

The formula for finding the rainwater run-off, in cubic metres a minute (m³/min), from a paved area can be found by using the formula

$$Q = \frac{APR}{6 \times 10^4}$$

where
- Q = run-off m³/min
- A = area to be drained in m²
- P = impermeability factor
- R = rainfall in mm/hr

Calculate the quantity of rainwater that will run off a paved area 100 m by 80 m in 5 minutes if the rainfall is 50 mm/hr and the impermeability is 0.75.

Answers and comments

Problem 3

Multiplication, division and addition are needed to answer this problem. I will first find the amount and cost of each pipe size required for one house, and will add them together and multiply by 12 twelve to obtain the final answer.

15 mm copper pipe
Amount required = 45 m
Number of 3 metre lengths = 45 ÷ 3 = 15
Price per length = £3.50, so 15 lengths at £3.50 a length = 15 × £3.50
= £52.50

22 mm copper pipe
Amount required = 24 m
Number of 3 metre lengths = 24 ÷ 3 = 8
Price per length = £5.00, so 8 lengths at £5.00 a length = 8 × £5.00
= £40.00

28 mm copper pipe
Amount required = 6 m
Number of 3 metre lengths = 6 ÷ 3 = 2
Price per length = £7.50, so 2 lengths at £7.50 a length = 2 × £7.50
= £15.00

Total cost of copper pipe for one house = £53.50 + £40.00 + £15.00
$$= £107.50$$
Total cost for twelve houses = £107.50 × 12 = £1,290.00

Problem 4

The length of the rafter can be found using Pythagoras' theory. It may help to think of the roof as a right-angled triangle as in Figure 12.6 with the line BC equalling half the span.

The line AB = the rise 2 m
The line BC = the span 5.2 m ÷ 2 = 2.6 m
The line AC = length of rafter (to be found).

$$AC^2 = AB^2 + BC^2$$
$$= 2^2 + 2.6^2$$
$$= 4 + 6.76$$
$$= 10.76$$

$$AC = \sqrt{10.76}$$
$$= 3.2802439 \text{ m}$$

Length of rafter to nearest millimetre = 3.280 m.
    The angle of the cut at the eaves (y) can be found as

$$y = 180° - 38° = 142°$$

The angle at the ridge (x) can be found as

$$x = 180° - \angle BAC$$
$$\angle BAC = 90° - 38° = 52°$$
$$x = 180° - 52° = 128°$$

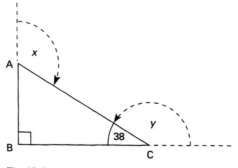

Fig. 12.6

Problem 5

There are two methods of finding the area of a shape that is a combination of a rectangle and a triangle, as is the rear wall of the bungalow and the profile of the concrete cills in the next problem. In this problem I shall divide the elevation into two areas, a rectangle and a triangle, but in Problem 6 I shall treat the shape as a whole. Either method is acceptable.

Area of rectangle $= 9 \text{ m} \times 4 \text{ m} = 36 \text{ m}^2$
Area of triangle $= (9 \text{ m} \times 2 \text{ m}) \div 2 = 9 \text{ m}^2$
Total area of rear elevation $= 36 \text{ m}^2 + 9 \text{ m}^2 = 45 \text{ m}^2$

From this the window area must be deducted.

Area of patio window $= 3 \text{ m} \times 2.1 \text{ m} = 6.3 \text{ m}^2$
Area of gable window $= 0.9 \text{ m} \times 1.1 \text{ m} = 0.99 \text{ m}^2$
Total area of windows $= 6.3 \text{ m}^2 + 0.99 \text{ m}^2 = 7.29 \text{ m}^2$
Total area of brickwork $= 45 \text{ m}^2 - 7.29 \text{ m}^2 = 37.71 \text{ m}^2$

Cost

37.71 m^2 at £10 a metre $= 37.1 \times 10 = £377.10$
37.71 m^2 at £2 a metre $= 37.1 \times 10 = £75.42$

Total charge for raking out and repointing $= £377.10 + £75.42 = £452.53$.

Problem 6

Area of profile of cills is equal to

$$0.410 \text{ m} \left(\frac{0.14 \text{ m} + 0.1 \text{ m}}{2} \right)$$

I have converted the measurements to metres

$$= 0.41 \text{ m} \times 0.12 \text{ m} = 0.0492 \text{ m}^2$$

The area of the end of each cill is 0.0492 m^2
Volume of cills is equal to

$2 \times 1.9 \text{ m} \times 0.0492 \text{ m}^2 = 0.18696 \text{ m}^3$
$6 \times 2.4 \text{ m} \times 0.0492 \text{ m}^2 = 0.70848 \text{ m}^3$
$6 \times 2.75 \text{ m} \times 0.0492 \text{ m}^2 = 0.8118 \text{ m}^3$
$4 \times 2.9 \text{ m} \times 0.0492 \text{ m}^2 = 0.57072 \text{ m}^3$
$6 \times 3.4 \text{ m} \times 0.0492 \text{ m}^2 = \underline{1.00368 \text{ m}^3}$
Total volume $\qquad\qquad = 3.28164 \text{ m}^3$

As the question requires the answer to the nearest half metre these figures could be rounded down to two decimal places

6% of $3.28164 \text{ m}^3 = 3.28164 \text{ m}^3 \times 0.06 = 0.1968984 \text{ m}^3$

Ready mixed concrete to be ordered is:

$3.28164 \text{ m}^3 + 0.1968984 \text{ m}^3 = 3.4785384 \text{ m}^3$

or, for practical reasons, 3.5 m^3.

Problem 7

Volume of earth to be excavated $= 40 \text{ m} \times 1 \text{ m} \times 0.75 \text{ m} = 30 \text{ m}^3$
Bulking of 30% of $30 \text{ m}^3 = 30 \text{ m}^3 \times 0.3 = 9 \text{ m}^3$
Volume of earth after bulking $= 39 \text{ m}^3$

Area left in trench after footings are completed = Area of trench − 48%

48% of 30 m³ = 30 m³ × 0.48 = 14.4 m³

Volume of earth to be removed from site = 39 m³ − 14.4 m³ = 24.6 m³.

Problem 8

A total of seven labourers were employed. Four labourers worked all day:

8 a.m. to 6 p.m. less 1 hour for lunch = 9 hours

Three labourers arrived 3 hours late:

11 a.m. to 6 p.m. less 1 hour for lunch = 6 hours

Combining their hours:

4 labourers × 9 hours = 36 man hours
3 labourers × 6 hours = 18 man hours

Total operative hours worked = 36 + 18 = 54

54 hours at £9 an hour = £486

Problem 9

Make r the subject of $V = \frac{1}{3}\pi r^2 h$. Transposing:

$$r = \sqrt{\frac{3V}{\pi h}} = \sqrt{\frac{3 \times 4.713 \text{ m}^3}{3.14 \times 2}}$$

$$= \sqrt{\frac{14.139}{6.264}} = \sqrt{2.25}$$

$$= 1.5$$

If you had any problems following the transposing of the formula, turn back to Chapter 11.

Diameter of cone = 1.5 m × 2 = 3 m

Problem 10

Substituting the letters in the formula $Q = \dfrac{APR}{6 \times 10^4}$ for the quantities given:

$$Q = \frac{100 \times 80 \times 0.75 \times 50}{6 \times 10^4}$$

$$= \frac{300,000}{6 \times 10^4} \quad \text{or} \quad \frac{3 \times 10^5}{6 \times 10^4} = \frac{10}{2} = 5 \text{ m}^3/\text{min}$$

Water run-off in 5 minutes

5 m³ × 5 minutes = 25 m³

Index